Contents

Biological Journal of the Linnean Society (1989), *37:* 1.

Evolution, ecology and environmental stress

The papers drawn together in this volume have all arisen out of a Symposium organized to celebrate the Bicentenary of the Linnean Society, held in the Rooms of the Society in June of 1988. The aim was to clarify the concept of stress and to consider its use as a probe of ecological systems and evolutionary processes. There was also an effort to consider the impact of stress at various levels of biological organization and to investigate the extent to which impacts at one level might be related to changes at others. Hence, apart from the first and last papers in this issue that give in turn general introductions and conclusions, the others are arranged, as they were at the Symposium, in order, from ecosystem, through population to individual, physiological and genetic levels of stress responses. Readers will judge for themselves how successful we have been in teasing out general principles and in finding links between levels.

The Symposium received financial support from the Linnean Society and the Royal Society. The organizing committee consisted of Brian Bayne (Plymouth Marine Laboratory), Phil Grime (Unit of Comparative Plant Ecology, University of Sheffield) and myself. Chairmen of sessions, who also assisted in refereeing manuscripts, were: Dr J. H. Steele (Woods Hole Oceanographic Institute), Professor W. Wieser (University of Innsbruck) and Professor R. J. Berry (University College London). Most of the burden of local organization was shouldered by Cdr. J. Fiddian-Green (Executive Secretary) and his staff and it is a pleasure to take this opportunity to thank them for that.

PETER CALOW
University of Sheffield

Biological Journal of the Linnean Society (1989), *37:* 3–17. With 2 figures

The stress debate: symptom of impending synthesis?

J. P. GRIME

Unit of Comparative Plant Ecology (N.E.R.C.), Department of Animal and Plant Sciences, The University, Sheffield S10 2TN

Opposition to use of the word 'stress' and other terms of wide ambit in ecology and evolutionary biology often signifies a commitment to programmes of detailed research and narrowed focus. This approach is justified where the objective is to analyse the demographic, genetic or biochemical mechanisms of contemporary populations. However, concepts employing 'stress' are vital to those forms of research that seek to obtain a broader perspective by recognizing ecological and evolutionary patterns with a high degree of repetition (*sensu* MacArthur, 1968 in R. C. Lewontin, *Population Biology & Evolution*). It is argued that there are recurrent types of evolutionary response to stress and that these provide important clues to the structure and dynamics of communities and ecosystems. Such patterns can assist our understanding of phenomena as widely divergent as the decline or expansion of populations within the British flora and fauna and the persistence of radionuclides in unproductive terrestrial ecosystems.

KEY WORDS:—Semantics – stress – life history – foraging – rarity – abundance – radionuclides – Chernobyl.

CONTENTS

INTRODUCTION

Arguments surrounding use of the word 'stress' in biology have two main origins. The first arises from the misuse in medicine and psychology of a term which, in its original context, in physics, described an externally applied force.

0024–4066/89/050003 + 15 $03.00/0

Here it will be maintained that the objectives of ecological and evolutionary research are best served by preserving this original meaning. Henceforward in this paper stress will be used to describe *external constraints limiting the rates of resource acquisition, growth or reproduction of organisms*. As in physics, the term will carry no implication with respect to the degree of deformation resulting in the stressed subject; this may vary from zero to complete inhibition and will depend critically upon the characteristics and evolutionary history of the subject and the nature, severity and periodicity of the stress. This establishes a useful commonality between stress analyses conducted upon living and non-living structures. As in the case of bars of metal of different alloys and thicknesses subjected to a given shearing stress, so organisms of different species, genotypes, phenotypes and maturity-states will differ in their responses to a particular environmental constraint.

A more profound and interesting source of debate about stress relates to the extent to which (in common with usages of certain other key words) adoption and abhorrence have often come to symbolize rival philosophies in the effort to devise generalizing principles in ecology. Science provides many earlier examples where conflict over terminology has acted both as surrogate for a more important and necessary debate and as a conservative reaction to the challenge of new ideas or novel synthesis (Kuhn, 1962; Mulkay, 1972). Against this background, therefore, the first objective in this paper will be to suggest some of the methodological and philosophical differences which lie behind current 'stress semantics'. The second and more specific objective will be to examine some of the opportunities for ecological analysis and predictions which follow from generalizations concerning stress and its evolutionary consequences.

THE STRESS DEBATE

Reference to publications which rely heavily upon the term 'stress' or comment upon its use in ecology (e.g. Grime, 1974, 1977; Harper, 1981, 1982; Pigott, 1980; Pugh, 1980; Grubb, 1980, 1985; Raven, 1981; Leps, Osbornova-Kosinova & Rejmanek, 1982; Cooke & Rayner, 1984; May & Seger, 1986; Larcher, 1987) reveal a sharp division of opinion. To a surprising extent, however, the matter at issue is not the meaning to be attached to 'stress', but rather the question as to *whether the term should be used at all*. Particularly vehement objections are lodged by Harper (1981, 1982) and Grubb (1985) in publications which also express reservations about the propriety of using certain other words, for example, 'strategy', for ecological purposes. A careful reading of Harper (1982) and Grubb (1985) reveals that both these authors object to 'stress' as part of a general concern to maintain a high degree of precision in ecological analysis and terminology. This suggests that conflict over the word 'stress' may have arisen for the comparatively trivial reason that different ecologists analyse nature at different scales (Allen & Starr, 1982) and with different degrees of precision. Robert MacArthur (1968) recognized this diversity of outlook as follows:

> Ecological patterns, about which we construct theories, are only interesting if they are repeated. They may be repeated in space or in time, and they may be repeated from species to species. A pattern which has all of these kinds of repetition is of special interest because of its generality, and yet

> these very general events are only seen by ecologists with rather blurred vision. The very sharp-sighted always find discrepancies and are able to say that there is no generality, only a spectrum of special cases. This diversity in outlook has proved useful in every science, but it is nowhere more marked than in ecology.

In the light of MacArthur's observation it is tempting to dismiss the stress debate merely as one of the many minor collisions likely to occur as the paths of research workers, intent on different objectives, cross each other. However, this cannot explain all of the debate; as the following quotations make plain, some of the argument is between ecologists who have similar objectives in view.

J. L. Harper:

> I worry . . . about the use of the word 'stress' in ecology. For the physicist it is a 'force per unit area'. Does the word bear any corresponding precision in ecology? I am also becoming unhappy with the usage of the word 'disturbance' in ecology. Again, it is not easy to see how it can be made operational. Is there some way in which a forest fire, the track of a bulldozer, a rabbit's burrow or a falling raindrop can be compared quantitatively as environmental forces? If we have to measure 'disturbance' by the response of the organisms, there is danger of a circular argument. Ideally, I suspect, we should aim to measure and compare the effects of various forces on individual fitness. Is this feasible?

J. P. Grime:

> In recent years, Professor Harper has contested the usefulness of most of the key words in my ecological vocabulary. . . . I readily concede that these terms have proved unnecessary in the increasingly specialized studies of plant demography and morphology. . . . With many other ecologists, I remain attached to a terminology that can assimilate information from all fields of ecological research and can play its part in the development of a general conceptual framework for plant and animal ecology.

Recorded discussion, pages 653–654 in
Philosophical Transactions of the
Royal Society of London Series B, 314 (1986).

'Stress' is thus embroiled in a wider and much more important debate about current priorities and methods in ecological research. From these exchanges it is clear that the wider debate refers not only to methodology and precision but also to the potential of existing concepts and data sources to support generalizing principles in ecology. 'Stress' is projected into the foreground of this discussion because, as with several other key words of wide ambit (e.g. disturbance, competition, strategy, niche, community and ecosystem) use implies the existence of pattern (*sensu* MacArthur, 1968) with the various kinds of repetition likely to permit ecological generalizations.

It is not the purpose of this paper to examine in detail the present scope for employing these various emotive words in a distillation of ecological theory; this subject has been explored in recent reviews (Southwood, 1988; Grime, 1988). Here attention will be confined to 'stress' and the advantages which may be gained by retaining its use in ecological analysis.

EVOLUTION AND STRESS RESPONSE

Organic evolution on earth emerged from a crucible of physical and chemical constraints and to the present day continues to be restricted in kind and quantity by the severe stresses operating over large portions of the planet's surface. Even in relatively hospitable environments, populations experience constraints arising from a multiplicity of factors related to weather, resource consumption by neighbours and release of toxins and pollutants. On first inspection, this presents a bewildering array of possibilities for natural selection and evolutionary response. The search for pattern is further complicated by the realization that the form and extent of response to contemporary challenge may be strongly conditioned by ancestral evolutionary attunements to earlier stresses. It is also evident that the exposure of an organism to stress and its scope for phenotypic or genetic adjustment will depend critically upon its nutritional mode, life history and breeding system.

Faced with this complexity, it is hardly surprising that many biologists have chosen to analyse stress responses through studies of contemporary evolution in amenable subjects and have decided that attempts at synthesis are *non grata* until the results of many more studies are available. Where the objective of this research is to identify the processes (demographic, genetic and biochemical) whereby particular populations respond to specific stresses, we may expect, in the short term at least, the development of many detailed lines of enquiry, each of necessity committed to a path of increasingly refined analysis and narrowed perspective.

ECOLOGY AND STRESS RESPONSE

Many of the pressing demands upon ecologists for scientific management of the biosphere involve prediction and analysis of stress responses at all levels from an individual organism to whole ecosystems. The animal and plant populations implicated in these problems are large in number and include some which, because of their size, longevity or mobility or the complexities of their life histories or interdependence with other organisms, are relatively intractable in field and laboratory. Such difficulties are not easy to address by programmes of intensive study of selected organisms and they have prompted an alternative research strategem. This relies upon the assertion that we can introduce priorities into ecological analysis by recognizing that evolutionary responses to stress fall into basic types which correspond to widely-recurrent ecological strategies. This is to suggest that patterns of evolutionary and ecological specialization are severely constrained such that all organisms, regardless of taxonomic or trophic affiliation, can be placed in a common framework of basic functional types. Each functional type is distinguishable as a set of traits generated and sustained by a characteristic pattern of natural selection, in which a most important element is the form, severity and periodicity of exposure to stress. Where this approach can be applied it may provide opportunities for extrapolation from the few intensively-studied organisms and habitats to the larger number which have been neglected.

Before considering further the potential of existing sources to support a typology of evolutionary responses to stress, it is essential to distinguish between

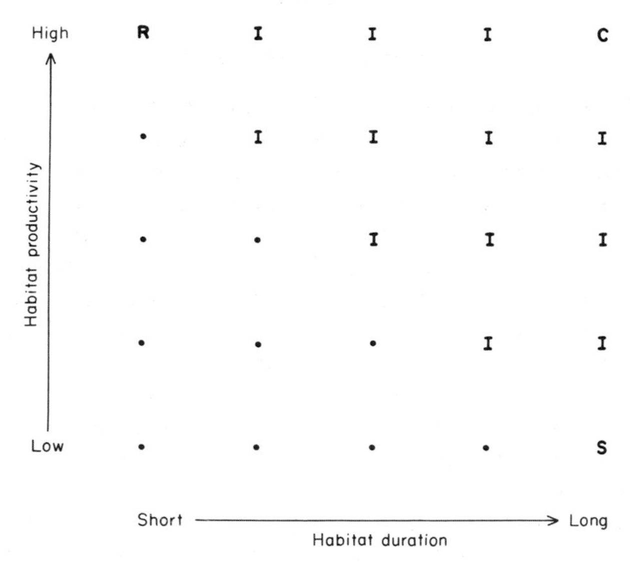

Figure 1. A matrix of evolutionary responses to stress, classified according to habitat duration and productivity. Key to responses: R–sustained reproduction; C–capitalism; S–stress-tolerance; I–intermediate; •–no viable habitat.

this ecological and essentially 'coarse-grained' dissection of the struggle for existence and the concern of some evolutionary biologists to establish beyond peradventure the origin and functional significance (if any) of individual traits (e.g. Gould & Lewontin, 1979). Few will dispute the need for caution in attributing an adaptive role to a particular trait observed in a single species or population. However, it is vital for the future of ecology that we should not become so cautious that the significance of recurring sets of traits engaging many, if not all, of the fundamental activities of organisms (resource capture, growth, defence and reproduction) passes unrecognized. It is important, however, to base conclusions on broadly-based comparisons; as explained by Clutton-Brock & Harvey (1979), "as more species are considered it becomes progressively more difficult to fit several adaptive hypotheses to the empirical facts".

EVOLUTIONARY RESPONSES TO STRESS IN THE ADULT (ESTABLISHED) PHASE

Variation in the mechanism of evolutionary attunement to stress in the adult phase of plant and animal life histories may be conveniently explored by reference to the matrix *habitat duration* × *habitat productivity* (Southwood, 1977; Grime, 1977). Although this matrix approaches the ultimate in simplification (Fig. 1), comparison with field reality suggests that it coincides with major axes of variation in life history and stress biology. A continuous triangular array including various intermediate forms of evolutionary response to stress is generated by the matrix but here it is necessary only to consider three extreme types of response—sustained reproduction, capitalism and stress-tolerance.

Sustained reproduction

In circumstances where high rates of resource capture are possible for short periods and the life of the individual is soon terminated by food exhaustion, predation or some other agency (e.g. garden weeds, bread moulds, fruit flies), natural selection is likely to favour those genotypes in which rapid growth and early reproduction increase the probability that sufficient offspring will be produced to maintain the population. It is also to be expected that plants and animals exploiting these conditions will evolve a characteristic phenotypic response to stress, particularly where the latter is predictably related to food exhaustion or other mechanisms which kill the adults. This response will be to sustain allocation to reproduction even at the cost of an increased rate of mortality in the adult population. Experiments in which various ephemeral organisms have been exposed to food shortage (e.g. Salisbury, 1942; Hickman, 1975; Calow & Woollhead, 1977; Boot, Raynal & Grime, 1986) support this prediction.

Capitalism

A second extremity within the triangular array of ecological specializations and attendant evolutionary responses to stress corresponds to circumstances which allow access to large concentrations of food materials over considerable periods of time (i.e. for several years at least). Examples here include the fast-growing clonal herbs, shrubs and trees of alluvial terraces and recently-abandoned arable fields, the extensive, perennial wood-rotting fungi that colonize timber and the wading birds that exploit the rich invertebrate fauna of estuarine mud. Despite the wide divergence in nutritional mode presented by these examples, all adhere to the same pattern of resource capture and utilization. In each case high rates of resource capture from an extensive area of habitat are maintained over a long period by a process of active foraging (Grime, 1977). Active foraging compensates for local patchiness within the resource mosaic, an inevitable consequence of the high rates of depletion caused by the organism and its competitors. In mobile animals, active foraging is achieved by constant locomotion and search, whereas the same process involves the attenuation and local proliferation of mycelia in fungi (Dowson, Rayner & Boddy, 1986; Rayner, Boddy & Dowson, 1987) and in root system and leaf canopies in vascular plants (Drew, 1975; Grime, Crick & Rincon, 1986; Crick & Grime, 1987; Slade & Hutchings, 1987a,b). For both animals and plants, however, active foraging is a costly activity involving high rates of re-investment of energy and other captured resources. An economic metaphor is appropriate here, with biological 'capitalism' identified as a distinctive evolutionary response to the dynamic spatial patterns of stress in rich but intensively exploited habitats and substrates.

Stress-tolerance

In marked contrast to the circumstances conducive to biological capitalism, there are numerous habitats in which, relative to the size of the individual, only low rates of capture of the limiting resources are possible. The slow-growing, comparatively long-lived cushion plants, mosses, beetles, molluscs and spiders of rock outcrops and other habitats undergoing primary succession illustrate this phenomenon, together with the various fungi that occur in lichens or exploit the more intractable residues in the terminal stages of fungal succession on decaying matter. Less obvious and more contentious examples include some large plants and animals (e.g. late successional trees, elephants, whales) which, by virtue of their size and longevity or the retention of captured resources, have the potential to enjoy a dominant status within mature, 'equilibrium' communities but are constrained by the scale of the demand which they themselves impose upon the resource base. In these various organisms, both large and small, survival and reproduction often depend crucially upon the capacity of both juveniles and adults to remain viable through long periods in which resource capture is limited by absolute shortage, intervention of climate or sequestration in other organisms. In conditions of chronic and pervading stress there is less scope for escape through phenology, phenotypic plasticity or locomotion. Greenslade (1972a,b, 1983), Grime (1977) and Whittaker & Goodman (1979) have put forward essentially similar lists of the attributes which they predict will be consistently associated with the evolutionary response to continuous stress. All predict that the potential growth-rates of individuals and populations will be relatively slow and will coincide with a marked uncoupling of resource intake from growth and reproductive activity, hence permitting the accumulation of considerable internal reserves. In strong contrast with the active spatial foraging of 'capitalist' organisms, it is predicted that resource capture in the most extreme stress-tolerators will be a more passive process in which brief, infrequent and unpredictable pulses of resource are intercepted. It is well-established that 'sit and wait' mechanisms of foraging are common among the animals of continuously hostile environments but it is only relatively recently (Crick & Grime, 1987; Campbell, 1988) that similar phenomena have been detected in bryophytes and flowering plants of unproductive habitats.

Stress-tolerance is also predicted to be associated with a characteristic reproductive response to unusually protracted or severe stress. Under such conditions it is expected that reproduction will be suspended or aborted, with benefit to the survivorship of the parent. Although this hypothesis has not been subjected to critical experimental tests, it is consistent with field evidence relating to a wide range of organisms exploiting harsh environments (French, Maza & Aschwanden, 1966; Keast, 1959; Bentley, 1966; Lloyd & Pigott, 1967; Billings & Mooney, 1968; Turner *et al.*, 1970; Jarvinen, 1986).

EVOLUTIONARY RESPONSES TO STRESS IN THE JUVENILE (REGENERATIVE) PHASE

From the contributions of many ecologists and evolutionary biologists, including Stebbins (1951, 1974), Sagar & Harper (1961), Wilbur, Tinkle & Collins (1974), Grubb (1977), Gill (1978) and Grime (1979), there has been gradual recognition of the advantage to be gained by separate analysis of the strategies exhibited by organisms during the regenerative phase of the life history. This acknowledges the principle that the microhabitats exploited by juveniles may be quite different from those experienced by the parents. In addition, many juveniles have characteristics, such as small size, high potential mobility, or capacity for dormancy which not only expose them to additional

TABLE 1. Five regenerative strategies of widespread occurrence in terrestrial vegetation

Strategy	Functional characteristics	Conditions under which strategy appears to enjoy a selective advantage
Vegetative expansion	New shoots vegetative in origin and remaining attached to parent plant until well established	Productive or unproductive habitats subject to low intensities of disturbance
Seasonal regeneration	Independent offspring (seeds or vegetative propagules) produced in a single cohort	Habitats subjected to seasonally predictable disturbance by climate or biotic factors
Persistent seed or spore bank	Viable but dormant seeds or spores present throughout the year; some persisting more than 12 months	Habitats subjected to temporally unpredictable disturbance
Numerous widely dispersed seeds or spores	Offspring numerous and exceedingly buoyant in air; widely dispersed and often of limited persistence	Habitats subjected to spatially unpredictable disturbance or relatively inaccessible (cliffs, walls, tree trunks, etc.)
Persistent juveniles	Offspring derived from an independent propagule but seedling or sporeling capable of long-term persistence in a juvenile state	Unproductive habitats subjected to low intensities of disturbance

hazards, but confer a rich potential for adaptive specialization.

For plants, theories of ecological specialization in the regenerative phase have been formalized as a set of five regenerative strategies (Table 1), supplemented in vascular plants by seed-bank models describing four major seasonal patterns of dormancy and persistence in the soil (Fig. 2). Detailed descriptions of these patterns, several of which have numerous close analogues in the regenerative biology of heterotrophs, are described in Grime (1979). Here it is necessary merely to point out that both the regenerative strategies and the seed-bank types are consistent in detail with the notion that natural selection has generated recurring sets of juvenile traits that confer either stress-tolerance or, more commonly, the capacity for stress-avoidance, by various mechanisms of dormancy or dispersal.

IMPLICATIONS

There is an urgent need for rigorous tests of current generalizations relating to stress and its evolutionary consequences; some of the procedures which are required for this task are examined by Grime, Hunt & Krzanowski (1987), Grime (1987) and Campbell (1988). A second challenge is to demonstrate that theories of evolutionary response to stress can inform our understanding of ecological processes such as those controlling succession, stability and diversity in plant and animal communities. With respect to plant communities, it is already possible to use the established and regenerative strategies as a tentative basis for prediction and interpretation of vegetation dynamics (van der Valk & Davis, 1976; Nobel & Slatyer, 1979; Grime, 1980, 1987; Leps *et al.*, 1982). As a further illustration, brief consideration is now given to two specific examples where generalizations concerning evolutionary responses to stress are of current interest.

Expansion and decline in British flora and fauna

As explained in Lousely (1953), Ratcliffe (1984) and Hodgson (1986a,b), changes in land use since about 1939 have wrought a profound and rapid modification of the British flora. These changes have been most conspicuous in Lowland England and have been associated with consistent patterns of both decline and expansion in particular native species (Perring & Walters, 1976). Using data recently compiled for all the common vascular plants of Inland Britain (Grime, Hodgson & Hunt, 1988), it is possible to examine the biological characteristics of species that are expanding or contracting. Of the 502 species examined, 66% appear to be decreasing and reference to the traits tabulated for these plants leaves little doubt that the majority are relatively stress-tolerant. This pattern may be explained as an inevitable consequence of the vulnerability of these comparatively slow-growing, long-lived plants of modest fecundity to the increased degree of habitat disturbance and to dominance by the more aggressive species encouraged by intensive cropping, increased mechanization and use of artificial fertilizers. As we might expect, the data also confirm the prediction that the majority of the species which are currently expanding in abundance are either ephemerals or capitalists.

Using the same data source, it is also possible to analyse the regenerative characteristics of expanding and declining components of the British flora. In Table 2, species, classified according to the four seed-bank types of Fig. 2, are compared with respect to current trends in their abundance. These data reveal a marked association between Seed-Bank Type III and the capacity for expansion under currently prevailing conditions. This finding is consistent with the hypothesis that Seed-Bank Type III has conferred ecological versatility in the circumstances created by the increasingly disruptive patterns of land use applied to the British landscape in recent decades. Such versatility is to be expected from Seed-Bank Type III because it combines the potential for immediate germination and rapid population expansion in disturbed habitats with the capacity for persistence as dormant seeds during conditions which are less favourable to seedling establishment. In Table 2, Seed-Bank Type II, which characteristically involves synchronous germination of seed populations in the early spring is correlated with declining abundance. This reflects the continuing loss of native broad-leaved woodland in which this type of seed bank predominates among both woody and herbaceous constituents.

TABLE 2. The proportion of increasing and decreasing species associated with different types of seed bank. Seed-banks I–IV are as described in Fig. 2

Seed bank	No. of spp.	Percentage increasing	Percentage decreasing	Increasing/ decreasing
I	33	27	52	0.53
II	39	18	66	0.32
III	48	50	33	1.50
IV	51	39	41	0.95

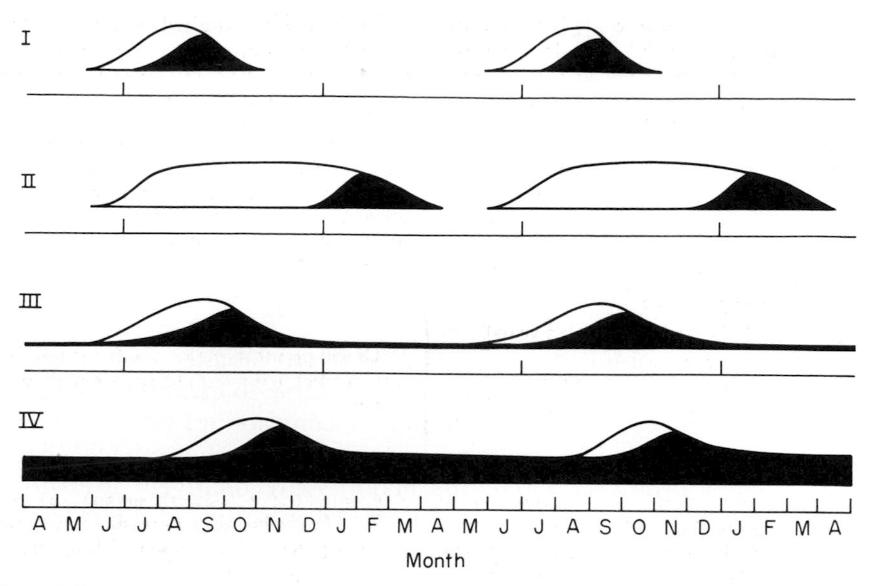

Figure 2. Representation of the four types of seed bank described by Thompson & Grime (1979). Shaded areas: seeds capable of germinating immediately after removal to suitable laboratory conditions. Unshaded areas: seeds viable but not capable of immediate germination. Type I, annual and perennial grasses of dry or disturbed habitats. Type II, annual and perennial herbs colonizing vegetation gaps in early spring. Type III, species mainly germinating in the autumn but maintaining a small persistent seed bank. Type IV, annual and perennial herbs and shrubs with large persistent seed banks.

In view of the consistent changes which are now occurring in the functional characteristics of the British flora, it would be most surprising if these were not found to be associated with impacts upon the fauna. Changes in vegetation have obvious effects upon the architecture of the habitats exploited by larger animals but much more profound impacts are to be expected from the massive increase in food quantity and quality which has attended the progressive displacement of a predominantly stress-tolerant and unpalatable (*sensu* Grime, 1979; Coley, 1983; Coley, Bryant & Chapin, 1985) phytomass by another consisting largely of short-lived, weakly-defended leaves and roots. Even allowing for the control measures applied to the pests of arable land, it seems likely that the botanical transformation and the continuing arrest of succession will have brought about a huge injection of tractable substrates and residues into the herbivore and decomposer systems. We should not be surprised therefore to note the apparently inexorable rise in British populations of birds such as starlings (*Sturnus vulgaris*) which exploit the invertebrate fauna of fertile agricultural ecosystems.

Persistence of radionuclides

The possibility that evolutionary responses to stress generate an essentially similar range of functional types in fungi, animals and green plants has interesting implications for the structure and dynamic properties of ecosystems. Recently Grime & Hodgson (1987) have pointed out that, in ecosystems of chronically low productivity, confinement of primary production to stress-tolerators is likely to have predictable consequences for other major components

TABLE 3. Some consistent differences between plants of fertile and infertile soils and their relevance to the persistence of ^{137}Cs in upland areas (simplified and adapted from Grime, 1979)

Plant characteristics	Species of fertile soils	Species of infertile soils	Implications for persistence of ^{137}Cs in upland areas
1. Life-form	Herbs, shrubs and trees	Lichens, bryophytes, herbs, shrubs and trees	Lichens and bryophytes intercept and absorb ^{137}Cs directly from rainwater through their above-ground surfaces. They also sequester metals for long periods, are unpalatable and decay slowly
2. Life-span of whole plant	Often short (<5 years)	Long (>5 years)	^{137}Cs will persist in plants of infertile soils because it is incorporated into potentially long-lived tissues
3. Life-span of individual leaves and roots	Short (<1 year)	Long (1–3 years)	
4. Leaf phenology	Well-defined peak of growth each spring and summer	Evergreens often showing no seasonal change in biomass	Presence of living foliage throughout the year in species of infertile soils will permit entrapment of ^{137}Cs regardless of the season of deposition
5. Maximum potential relative growth-rate	Rapid	Slow	Slow growth of plants of infertile soils will mean lower rates of tissue turnover and tendency to retain ^{137}Cs for long periods
6. Uptake of mineral nutrients and other ions	Strongly seasonal; and mostly in spring and summer	Opportunistic capable of occurring at most times of the year	Presence of functional roots throughout the year will allow accumulation of ^{137}Cs from mineralization pulses regardless of their season of release into the soil solution
7. Mycorrhizal infection of root system	Light	Heavy	Mycorrhizal fungi are the effective absorbing surface for the root systems of plants of infertile soils and have a very high affinity for metals such as ^{137}Cs. They form an effective network throughout the surface soil and produce residues which are resistant to decay (see 9 and 10)
8. Storage of mineral nutrients	Most mineral nutrients are rapidly incorporated growth but a proportion is stored and forms the capital for expansion of growth in the following growing season. Little internal recycling from old leaves to new	Storage systems in leaves, stems and/or roots. Some recycling of minerals from old leaves to new	Because of the weak coupling between mineral uptake and utilization in growth, ^{137}Cs will tend to be retained in the plant biomass. Some internal recycling of ^{137}Cs is likely to occur, further retaining ^{137}Cs in the living tissues
9. Palatability	High	Low	Low palatability dictates lower density of sheep in upland pastures and results in a situation where much of the ^{137}Cs is likely to reside in physically repellant plant tissues which tend to be avoided by the animals except in winter when the supply of more palatable leaves is minimal

TABLE 3 — *continued*

Plant characteristics	Species of fertile soils	Species of infertile soils	Implications for persistence of ^{137}Cs in upland areas
10. Rate of litter decomposition	High	Low	The leaf toughness which protects the canopy of slow-growing plants of infertile pastures from heavy defoliation by sheep and invertebrates, remains operational when the leaves die and fall onto the ground surface. In consequence, there is a reduced rate of decay and ^{137}Cs would be expected to be recycled back into the plants at a slow rate. Mycorrhizal roots (especially those associated with ericaceous plants, e.g. heather) tend to be very resistant to decay and will retain ^{137}Cs on the melanized cell walls of the fungal residues

of the biomass. Stress-tolerant plants produce tissues which are potentially long-lived and are strongly protected against herbivores (Chadwick, 1960; Gimingham, 1960; Grime, MacPherson-Stewart & Dearman, 1968; Williamson, 1976; Al-Mufti *et al.*, 1977; Sydes, 1984; Profitt, 1985). These defences are usually physical in nature (Coley, 1983; Coley *et al.*, 1985) and remain operational against decomposing organisms when they are assimilated into the litter component. On this basis, therefore, it seems reasonable to propose that the resistant nature of the living and dead tissues of stress-tolerant plants will often dictate parallels in life history, economics of resource capture and population dynamics between the *dominant organisms* of the various trophic components of an unproductive ecosystem.

In an unproductive ecosystem, it is also predictable that 'slow dynamics' at various scales of resolution, from cell turnover to exchanges between trophic compartments, will result in long residence times for chemically-stable pollutants. A recent example of this phenomenon is the persistence in areas of unproductive upland pastures in Britain and other sites in Northern Europe of radiocaesium, originating from the Chernobyl accident. Table 3 attempts to explain in greater detail the way in which the slow dynamics of unproductive ecosystems might lead to greater persistence by ^{137}Cs than would be predicted from models based on productive conditions.

CONCLUSIONS

The stress debate is a signal of divided opinion concerning the future conduct of ecological research. The disparate nature of ecological investigations and the fragmentary state of ecological understanding discourage synthesis and encourage a swift return to the many specialist enclaves where interim objectives can be pursued with precision and rigour. A consequence of this tendency is the continuing difficulty of translating specialist knowledge into the broad guidelines needed for ecological synthesis and the coarse control of communities and ecosystems.

The neo-Darwinian synthesis has brought full realization of the extent to which natural selection acting upon genetic variability has generated the current diversity of life and continues to make contemporary populations responsive to change. This guarantees that evolutionary concepts will continue to underpin much ecological research. However, contrary to the view of Harper (1982), recognition of natural selection as the proximal driving force does not mean that the detailed analysis of current population processes and genetic variation provides the key to general ecological enlightenment. Following MacArthur (1968), it seems necessary from time to time to bring into focus the most widely-recurring patterns of organic evolution. These appear to reflect fundamental constraints of habitat and organism which channel evolution into predictable paths. A current challenge is to assess the extent to which recognition of these patterns provides the essential clues to community and ecosystem structure.

ACKNOWLEDGEMENT

It is a pleasure to acknowledge the many different contributions of colleagues at UCPE to the field and laboratory studies on which this paper relies. This research was fully supported by the Natural Environment Research Council.

REFERENCES

ALLEN, T. F. H. & STARR, T. F., 1982. *Hierarchy: Perspectives for Ecological Complexity.* Chicago: University of Chicago Press.

AL-MUFTI, M. M., SYDES, C. L., FURNESS, S. B., GRIME, J. P. & BAND, S. R., 1977. A quantitative analysis of shoot phenology and dominance in herbaceous vegetation. *Journal of Ecology, 65:* 759–792.

BENTLEY, P. J., 1966. Adaptations of amphibia to arid environments. *Science, 152:* 619–623.

BILLINGS, W. D. & MOONEY, H. A., 1968. The ecology of arctic and alpine plants. *Biological Reviews, 43:* 481–529.

BOOT, R., RAYNAL, D. J. & GRIME, J. P., 1986. A comparative study of the influence of drought stress on flowering in *Urtica dioica* and *U. urens. Journal of Ecology, 74:* 485–495.

CALOW, P. & WOOLLHEAD, A. S., 1977. The relationship between ration, reproductive effort, and age-specific mortality in the evolution of life-history strategies—some observations on freshwater triclads. *Journal of Animal Ecology, 46:* 765–781.

CAMPBELL, B. D., 1988. *Experimental tests of C-S-R strategy theory.* Unpublished Ph.D. thesis, University of Sheffield.

CHADWICK, M. J., 1960. Biological Flora of the British Isles. *Nardus stricta* L. *Journal of Ecology, 48:* 255–267.

CHAPIN, F. S., 1980. The mineral nutrition of wild plants. *Annual Review of Ecology and Systematics, 11:* 233–260.

CLUTTON-BROCK, T. H. & HARVEY, P. H., 1979. Comparison and adaptation. *Proceedings of the Royal Society of London, Series B, 205:* 547–565.

COLEY, P. D., 1983. Herbivory and defensive characteristics of tree species in a lowland tropical forest. *Ecological Monographs, 53:* 209–233.

COLEY, P. D., BRYANT, J. P. & CHAPIN, F. S., 1985. Resource availability and plant antiherbivore defence. *Science, 230:* 895–899.

COOKE, R. C. & RAYNER, A. D. M., 1984. *Ecology of Saprotrophic Fungi.* London: Longman.

CRICK, J. C. & GRIME, J. P., 1987. Morphological plasticity and mineral nutrient capture in two herbaceous species of contrasted ecology. *New Phytologist, 107:* 403–414.

DOWSON, C. G., RAYNER, A. D. M. & BODDY, L., 1986. Outgrowth patterns of mycelial cord-forming Basidiomycetes from and between woody resource units in soil. *Journal of General Microbiology, 132:* 203–211.

DREW, M. C., 1975. Comparison of the effects of a localized supply of phosphate, nitrate, ammonium and potassium on the growth of the seminal root system, and the shoot, in barley. *New Phytologist, 75:* 479–490.

FRENCH, N. R., MAZA, B. G. & ASCHWANDEN, A. P., 1966. Periodicity of desert rodent activity. *Science, 154:* 1194–1195.

GILL, D. E., 1978. On selection at high population density. *Ecology, 59:* 1289–1291.

GIMINGHAM, C. H., 1960. Biological Flora of the British Isles. *Calluna vulgaris* L. Hull. *Journal of Ecology, 48:* 455–483.

GOULD, S. J. & LEWONTIN, R. C., 1979. The spandrels of San Marco and the Panglossian paradigm: a critique of the adaptationist programme. *Proceedings of the Royal Society of London, Series B, 205:* 581–598.

GREENSLADE, P. J. M., 1972a. Distribution patterns of *Priochirus* species Coeleoptera: Staphylindae in the Solomon Islands. *Evolution, 26:* 130–142.

GREENSLADE, P. J. M., 1972b. Evolution in the staphylinid genus *Priochirus* Coeleoptera. *Evolution, 26:* 203–220.

GREENSLADE, P. J. M., 1983. Adversity selection and the habitat templet. *American Naturalist, 122:* 352–365.
GRIME, J. P., 1974. Vegetation classification by reference to strategies. *Nature, 250:* 26–31.
GRIME, J. P., 1977. Evidence for the existence of three primary strategies in plants and its relevance to ecological and evolutionary theory. *American Naturalist, 111:* 1169–1194.
GRIME, J. P., 1979. *Plant Strategies and Vegetation Processes.* Chichester: Wiley & Sons.
GRIME, J. P., 1980. An ecological approach to management. In I. H. Rorison & R. Hunt, *Amenity Grassland: an Ecological Perspective:* 13–55. Chichester: Wiley & Sons.
GRIME, J. P., 1986. The circumstances and characteristics of spoil colonization within a local flora. *Philosophical Transactions of the Royal Society of London, Series B, 314:* 637–654.
GRIME, J. P., 1987. Dominant and subordinate components of plant communities: implications for succession, stability and diversity. In A. Gray, P. Edwards & M. Crawley, *Colonisation, Succession and Stability:* 413–428. Oxford: Blackwell Scientific Publications.
GRIME, J. P., 1988. The C-S-R model of primary plant strategies – origins, implications and tests. In L. D. Gottlieb & K. S. Jain, *Plant Evolutionary Biology:* 371–393. London: Chapman & Hall.
GRIME, J. P. & HODGSON, J. G., 1987. Botanical contributions to contemporary ecological theory. *New Phytologist, 106 Suppl.:* 283–295.
GRIME, J. P., CRICK, J. C. & RINCON, E., 1986. The ecological significance of plasticity. In D. H. Jennings & A. J. Trewavas, *Plasticity in Plants.* Symposia of the Society for Experimental Biology XL: 5–29. Cambridge: Society for Experimental Biology.
GRIME, J. P., HODGSON, J. & HUNT, R., 1988. *Comparative Plant Ecology: a Functional Approach to Common British Species.* London: Unwin Hyman.
GRIME, J. P., HUNT, R. & KRZANOWSKI, W. J., 1987. Evolutionary physiological ecology of plants. In P. Calow, *Evolutionary Physiological Ecology:* 105–126. Cambridge: Cambridge University Press.
GRIME, J. P., MACPHERSON-STEWART, S. F. & DEARMAN, R. S., 1968. An investigation of leaf palatability using the snail *Cepaea nemoralis* L. *Journal of Ecology, 56:* 405–420.
GRUBB, P. J., 1977. The maintenance of species-richness in plant communities: the importance of the regenerative niche. *Biological Reviews, 52:* 107–145.
GRUBB, P. J., 1980. Review: *Plant Strategies and Vegetation Processes. Journal of Ecology, 86:* 123–124.
GRUBB, P. J., 1985. Plant populations and vegetation in relation to habitat, disturbance and competition: problems of generalization. In J. White, *The Population Structure of Vegetation:* 595–621. Dordrecht: Junk.
HARPER, J. L., 1981. Review: *Environmental Physiology of Plants. Nature, 295:* 470.
HARPER, J. L., 1982. After description. In E. I. Newman, *The Plant Community as a Working Mechanism:* 11–25. Special Publication No. 1, British Ecological Society. Oxford: Blackwell Scientific Publications.
HICKMAN, J. C., 1975. Environmental unpredictability and plastic energy allocation strategies in the annual *Polygonum cascadense* Polygonaceae. *Journal of Ecology, 63:* 689–701.
HODGSON, J. G., 1986a. Commonness and rarity in plants with special reference to the Sheffield Flora. Part I. The identity, distribution and habitat characteristics of the common and rare species. *Biological Conservation, 36:* 199–252.
HODGSON, J. G., 1986b. Commonness and rarity in plants with special reference to the Sheffield Flora. Part II. The relative importance of climate, soils and land use. *Biological Conservation, 36:* 253–274.
JARVINEN, A., 1986. Clutch size of passerines in harsh environments. *Oikos, 46:* 365–371.
KEAST, A., 1959. Australian birds: their zoogeography and adaptation to an arid continent. In A. Keast, R. L. Crocker & C. S. Christian, *Biogeography and Ecology in Australia:* 89–114. The Hague: Junk.
KUHN, T. S., 1962. *The Structure of Scientific Revolutions.* Chicago: Chicago University Press.
LARCHER, W., 1987. Stress bei pflanzen. *Naturwissenschaften, 74:* 158–167.
LEPS, J., OSBORNOVA-KOSINOVA, J. & REJMANEK, K., 1982. Community stability, complexity and species life-history strategies. *Vegetatio, 50:* 53–63.
LLOYD, P. S. & PIGOTT, C. D., 1967. The influence of soil conditions in the course of succession on the chalk of southern England. *Journal of Ecology, 55:* 137–146.
LOUSLEY, J. E. (Ed.), 1953. *The Changing Flora of Britain.* Arbroath: Buncle.
MACARTHUR, R. H., 1968. The theory of the niche. In R. C. Lewontin, *Population Biology and Evolution:* 159–176. New York: Syracuse University Press, Syracuse.
MAY, R. M. & SEGER, J., 1986. Ideas in ecology. *American Scientist, 74:* 256–267.
MULKAY, M. J., 1972. *The Social Process of Innovation.* London: Macmillan.
NOBLE, I. R. & SLATYER, R. O., 1979. The use of vital attributes to predict successional changes in plant communities subject to recurrent disturbances. *Vegetatio, 43:* 5–21.
PERRING, F. H. & WALTERS, S. M., 1976. *Atlas of the British Flora.* 2nd edn. Published for the Botanical Society of the British Isles. Wakefield: E. P. Publishing.
PIGOTT, C. D., 1980. Review: *Plant Strategies and Vegetation Processes. Journal of Ecology, 68:* 704–706.
PROFITT, G. W. H., 1985. *The biology and ecology of Purple-moor Grass* Molinia caerulea L. *Moench. with special reference to the root system.* Unpublished Ph.D. Thesis, University of Aberystwyth.
PUGH, G. J. F., 1980. Strategies in fungal ecology. *Transactions of the British Mycorrhizal Society, 75:* 1–14.
RATCLIFFE, D. A., 1984. Post-mediaeval and recent changes in British vegetation: the culminating of human influence. *New Phytologist, 98:* 73–100.
RAVEN, J. A., 1981. Nutritional strategies of submerged benthic plants; the acquisition of C, N and P by rhizophytes and hapophytes. *New Phytologist, 88:* 1–30.

RAYNER, A. D. M., BODDY, L. & DOWSON, C. G., 1987. Genetic interactions and developmental versatility during establishment of decomposer Basidiomycetes in wood and tree litter. In M. Fletcher, T. R. G. Gray & J. G. Jones, *Ecology of Microbial Communities:* 83–123. Cambridge: Cambridge University Press.

SAGAR, G. R. & HARPER, J. L., 1961. Controlled interference with natural populations of *Plantago lanceolata, P. major* and *P. media. Weed Research, 1:* 163–176.

SALISBURY, E. J., 1942. *The Reproductive Capacity of Plants.* London: Bell.

SLADE, A. J. & HUTCHINGS, M. J., 1987a. The effect of nutrient availability on foraging in the clonal herb *Glechoma hederacea. Journal of Ecology, 75:* 95–112.

SLADE, A. J. & HUTCHINGS, M. J., 1987b. Clonal integration and plasticity in foraging behaviour in *Glechome hederacea. Journal of Ecology, 75:* 1023–1036.

SOUTHWOOD, T. R. E., 1977. Habitat: the templet for ecological strategies? *Journal of Animal Ecology, 46:* 337–365.

SOUTHWOOD, T. R. E., 1988. Tactics, strategies and templets. *Oikos, 52:* 3–18.

STEBBINS, G. L., 1951. Natural selection and the differentiation of angiosperm families. *Evolution, 5:* 299–324.

STEBBINS, G. L., 1974. *Flowering Plants: Evolution Above the Species Level.* London: Edward Arnold.

SYDES, C. L., 1984. A comparative study of leaf demography in limestone grassland. *Journal of Ecology, 72:* 331–345.

THOMPSON, K. & GRIME, J. P., 1979. Seasonal variation in the seed banks of herbaceous species in ten contrasting habitats. *Journal of Ecology, 67:* 893–921.

TURNER, F. B., HODDENBACH, G. A., MEDICA, P. A. & LANNOMS, J. R., 1970. The demography of the lizard, *Uta stansburiana* Baird and Birard, in southern Nevada. *Journal of Animal Ecology, 39:* 505–519.

VAN DER VALK, A. G. & DAVIS, C. B., 1976. The seed banks of prairie glacial marshes. *Canadian Journal of Botany, 54:* 1832–1838.

WHITTAKER, R. H. & GOODMAN, D., 1979. Classifying species according to their demographic strategy. I. Population fluctuations and environmental heterogeneity. *American Naturalist, 113:* 185–200.

WILBUR, H. M., TINKLE, D. W. & COLLINS, J. P., 1974. Environmental certainty, trophic level and resource availability in life history evolution. *American Naturalist, 108:* 805–817.

WILLIAMSON, P., 1976. Above-ground primary production of chalk grassland allowing for leaf death. *Journal of Ecology, 64:* 1059–1075.

Biological Journal of the Linnean Society (1989), *37:* 19–32. With 5 figures

Effects of environmental stress on species rich assemblages

JOHN S. GRAY

Department of Marine Biology, Biology Institute, University of Oslo, P.B. 1064. 0316 Blindern, Oslo 3, Norway

Selye's widely used model of responses of individual organisms to a stressor (1973, *American Science, 61:* 692–699) is not appropriate for describing effects at the population or community level. At the ecosystem level a number of functional responses have been suggested by Rapport, Regier & Hutchinson (1985, *American Naturalist, 125:* 617–640) but detailed analysis shows that, in general, functional responses are not sensitive to the early detection of impending ecosystem damage.

Three clear changes in community structure occur in response to stressors. These are reduction in diversity, retrogression to dominance by opportunist species and reduction in mean size of the dominating species. Statistically significant reductions in diversity occur rather late in the sequence of increased stressor impact. The first stages of impact are clearly shown by moderately common species yet most attention has concentrated on the common species. Species which dominate in heavily stressed habitats are often species complexes and the possible genetic mechanisms causing this are considered.

Whilst changes in the mean size of the dominant organisms can be shown in experiments there is no clear evidence that recorded reductions in the size of North Atlantic and North Sea plankton are induced by man-made stressors.

KEY WORDS:—Stress – diversity – dominance – size.

CONTENTS

INTRODUCTION

Investigating effects of stress on multi-species assemblages usually involves examining patterns of presence and absence of species and/or changes in abundance of sets of species. Presence/absence patterns imply that effects of a stressor (or stressors) have been severe and led to eradication of one or more species in a local area. Yet a major goal of monitoring the 'health' of an area is the early detection of effects of stressor(s). Ideally, detection should be at the level of the individual where performance is altered but death does not result. The range of species on which we have this kind of detailed knowledge is extremely narrow. The extrapolation of effects at the individual level to effects at

0024–4066/89/050019 + 14 $03.00/0

the population level is not an easy exercise (but see Underwood, this volume). Yet unless stressors can be shown as having effects at the population level these are unlikely to be ecologically significant (McIntyre & Pearce, 1981).

Here I shall use primarily marine examples to illustrate these points. One of the most striking aspects of marine biology is the high taxonomic diversity found in the plankton and benthos. Examination of effects of stressors on marine assemblages, therefore, has to take account of this high diversity.

However, there is an additional complicating factor specific to the marine environment in that many species produce planktonic larvae. Natural mortality on larval phases is usually high and thus detection of effects of stressors at the population level against a background of high natural mortality is difficult. In addition selective forces on the planktonic phase are much different to those affecting the adult. Recently, Roughgarden, Iwasawa & Baxter (1985) have made significant advances in understanding how natural processes affect recruitment of larvae to populations in *Balanus* populations and thereby predicting consequences at the population level.

It is, nevertheless, a difficult step to move from a study of one or a few species to predicting effects on complete assemblages. An alternative approach is to examine whole assemblages along gradients of stressors and hope to unravel common response patterns. However, there have been few attempts to compare the patterns obtained in response to different stressors or to look at the evolutionary significance of the data obtained. In this paper I shall review a number of examples of how stressors affect different marine benthic assemblages and attempt to relate these data to a theoretical and evolutionary base.

SELYE'S GENERAL ADAPTATION SYNDROME

Selye (1973) suggested a model of the response of an organism to a stressor. Figure 1 shows a modification of this model. A single application of a stressor (1) may lead to a measurable response within the organism (the 'alarm' reaction) followed by compensation to the stressor and finally return to normal activity. Alternatively if the application of the stressor is continuous or at an increased level (2) where the organism cannot cope, ultimately death results. Within individuals typical 'alarm reaction' responses are inhibition of cholinesterase production in response to organophosphorus pesticides in fish (Verma *et al.*, 1981) and birds (Busby, Pearce & Garrity, 1981; Ludke, Hill & Dieter, 1975), activation of mixed function oxidation enzymes in response to the presence of PAHs (Stegeman, 1980; Stegeman, Woodin & Goksøyr, 1988; Addison, 1984, 1988), or metallothionein production in response to heavy metals (Brown *et al.*, 1977; Brown & Parsons, 1978), or reduction in 'scope for growth' in the marine bivalve *Mytilus edulis* in response to a variety of stressors (Bayne, 1980; Bayne and Worral, 1980; Widdows & Johnson, 1988). Such responses provide early warnings that continued application of the stressor can lead to death. The above responses are measured at the individual level and few attempts have been made to infer changes at the population level. However, Bayne & Worral (1980) have shown that the 'scope for growth' response is due to diversion of energetic resources from growth to maintenance of gamete production. Further application of the stressor leads to reduced fecundity and presumed consequences at the population level (see Koehn and Bayne, this volume).

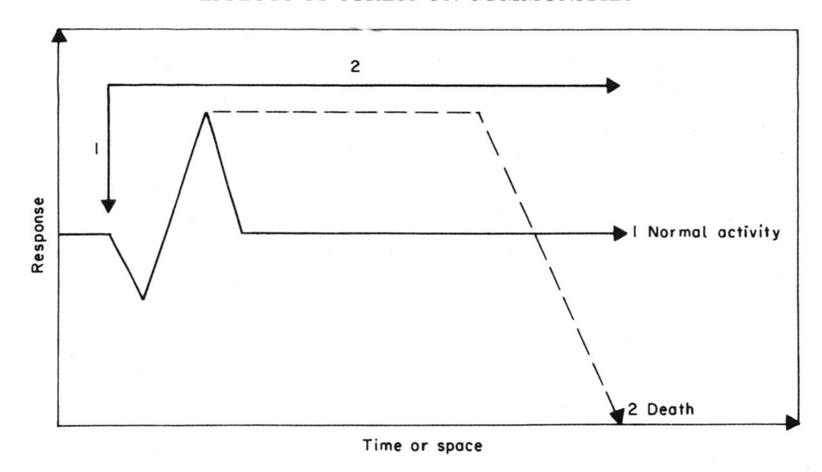

Figure 1. Modification of Selye's General Adaptation Syndrome. Arrows show application of stressor. Continuous line response to single application, broken line response to continuous application or increased level of stressor.

Selye's model applies reasonably well to responses at the individual level but does not describe significant effects at the population level. Above the population level a variety of different responses to the effects of stressors have been suggested.

EFFECTS OF STRESSORS ABOVE THE POPULATION LEVEL

Rapport, Regier & Hutchinson (1985) and Schindler (1987) have reviewed effects of stressors on ecosystems. Table 1 lists the responses that have been suggested.

The general functional responses of ecosystems to stressors such as production, decomposition and nutrient cycling, are varied. For example, primary production may increase (e.g. plankton in the Great Lakes in response to increased nutrient loading; Regier & Hartman, 1973) or decrease (e.g. pine trees in response to air pollution; Williams, 1980). Similar examples can be found for decomposition and nutrient cycling. Schindler (1987), based on review of many case histories, suggested that because they have feedback "monitoring ecosystem functions would be a poor approach to detecting early signs of impending ecosytem damage". In additon, an altered rate of a functional response must be measured by comparison with levels in an unstressed system. Often the unstressed system is poorly known so that it is not possible to detect a statistically significant change.

TABLE 1. Ecosystem stress symptoms (from Rapport, Regier & Hutchinson, 1985; Schindler, 1987)

1. Altered primary production
2. Altered rates of decomposition
3. Altered rates of nutrient cycling
4. Increased frequency of disease
5. Changed amplitude of fluctuations
6. Reduced diversity
7. Retrogression to opportunist species
8. Reduction in size

For example, primary production has been measured at fixed sites in the Baltic Sea for decades and this area has been subjected to increased nutrient enrichment. Yet it is not possible to say that annual primary production has increased (Wulff, 1979). One of the major problems is that to obtain reliable estimates of annual primary production one must sample very intensively, particularly over the period of peak production. This has rarely been done. Thus one can conclude that the first three suggested responses in Table 1 are not likely to be reliable indices of effects of stressors at the community level.

Likewise, although increased frequency of disease has been suggested as an indicator of stressed systems (Rapport *et al.,* 1985), recent reviews, for example on fish diseases in polluted marine areas (Moller & Anders, 1986; McVikar, Bruno & Fraser, 1988), suggest that the data are equivocal and there is no clear trend.

Regier & Hartman (1973) have suggested that the effects of stressors such as eutrophication and fishing led to destabilized population buffering mechanisms on commercial fisheries in the Great Lakes and increased amplitude of fluctuations. There are, however, few other documented cases.

The remaining three responses, reduction in diversity, retrogression to opportunist species and reduction in size, are all well-documented and will be treated in more detail.

REDUCTION IN DIVERSITY

Figure 2A shows reduction in the diversity of benthic macrofauna along a gradient of increasing hydrocarbon concentration in sediment approaching the Statfjord C oil platform in the North Sea. There are clear effects above 100 p.p.m. Figure 2B shows data for the benthic macrofauna of an area subjected to discharge from a titanium dioxide factory where the waste material is discharged into the sea. Titanium dioxide is not toxic and the concentration of TiO_2 in the sediment rather reflects the gradient of sedimentation rates of inert particles. Again clear effects on diversity are seen. Many similar examples of reduction in diversity in polluted areas are known; for example Patrick's (1972) data on diatoms in fresh-water streams, for marine organic enrichment gradients

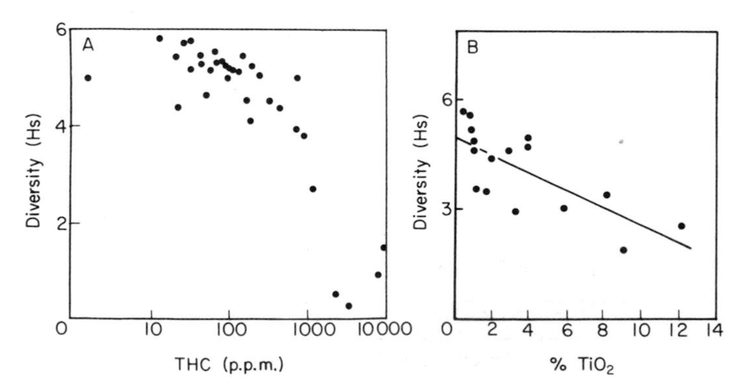

Figure 2. A, Reduction of diversity of benthic macrofauna along a gradient of increasing hydrocarbon concentration at the Statfjord C platform, Norway 1985. THC = Total hydrocarbon concentration (Data from Olsgaard *et al.,* unpublished). B, Reduction in diversity of benthic macrofauna with increasing concentration of titanium dioxide waste from Jøssingfjord, Norway 1985 (Data from Olsgaard *et al.* unpublished).

(Gray, 1979), for air pollution and radiation effects on trees (Woodwell, 1967) and see Rapport *et al.* (1985).

Although these effects are clear, there is little doubt that significant reductions in diversity are found rather late in the sequence of responses to stressors. Diversity integrates the number of species in the assemblage and the proportion of individuals per species. Examination of the components of diversity rather than a composite index leads us to the second well-documented change in ecosystems, the retrogression to dominance by opportunist species.

RETROGRESSION TO DOMINANCE BY OPPORTUNIST SPECIES

Figure 3 shows data for the number of species, abundance and biomass along some marine organic enrichment gradients. From these and many other data sets Pearson & Rosenberg (1978) have produced a general model of effects of organic

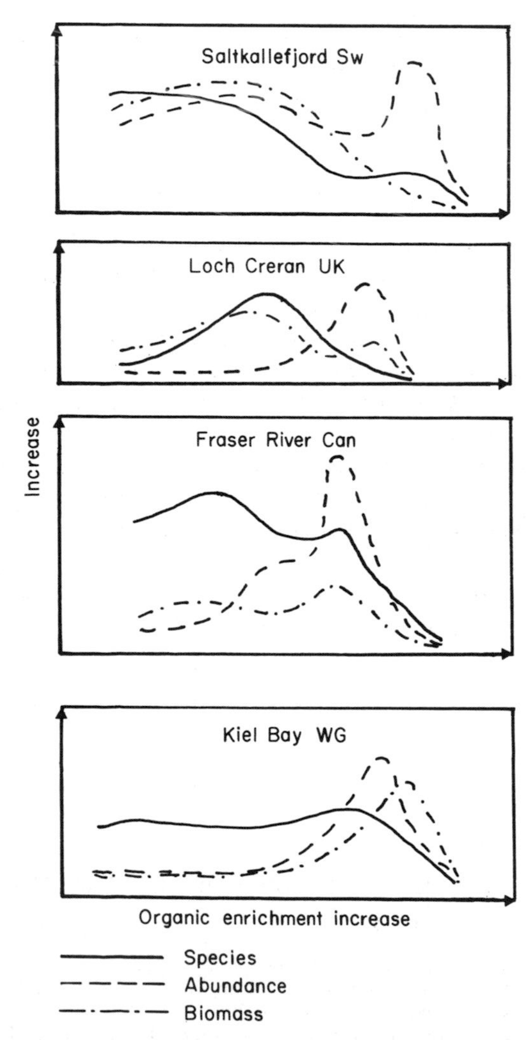

Figure 3. Effects of organic enrichment on marine benthic macrofauna. Data from Pearson & Rosenberg, 1978.

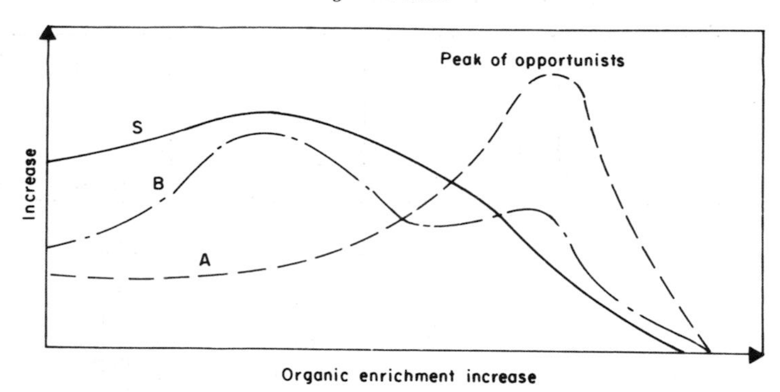

Figure 4. Generalized model of effects of organic enrichment on species (S), abundance (A) and biomass (B) in marine macrobenthos. After Pearson & Rosenberg (1978).

enrichment on marine benthos (Fig. 4). The model predicts that the first changes are a slight increase in number of species and a rapid increase in biomass, followed by a fall in species number and biomass. Abundance increases where the number of species starts to fall and at maximal abundance there is a second peak in biomass before both abundance and biomass fall sharply with increasing stressor load. Maximal abundance occurs at the 'peak of opportunists' where there is dominance by a few extremely common opportunist species. Yet the dominance by the opportunists is almost the last stage in the retrogression of the system. Comparing the initial unstressed to the stressed condition may give insights into important patterns of change.

If instead of total abundance and total biomass of the assemblage, one compares the rank of abundance and biomass among species, a distinction is seen between the pattern obtained in unpolluted and polluted conditions (Warwick, 1986). This abundance-biomass comparison (ABC curve) has been applied to a number of cases and suggested as a universal method for distinguishing polluted from unpolluted conditions (see Warwick, Pearson & Ruswahyuni, 1987). Beukema (1988) has recently shown that data from intertidal sites in Holland fit the 'polluted' pattern although the sites were not polluted. The intertidal area is subjected to a greater range of environmental variables than subtidal habitats, so perhaps the model distinguishes sites by stress rather than between polluted and unpolluted sites.

The detection of the first stages of stressor impact on assemblages is one major goal of ecological research. From the Pearson and Rosenberg model, species number and abundance change before the peak of opportunists is reached. Plots of number of individuals per species against number of species for different sites along a stressor gradient, show clear trends. Figure 5 shows data along a transect across a sewage dumping ground (Gray & Pearson, 1982). The unstressed sites (1,2,3 and 9,10) show steep curves with many rare species and few common species whereas the stressed sites show few rare species and dominance by a few highly common species. From this pattern (and many other examples), a general model was produced (Gray & Pearson, 1982) which suggests that the first phase of response to a stressor is that some rare species decrease in abundance and are eliminated from the community and moderately common species increase in abundance. Finally, in response to severe stressor impact, many rare species are

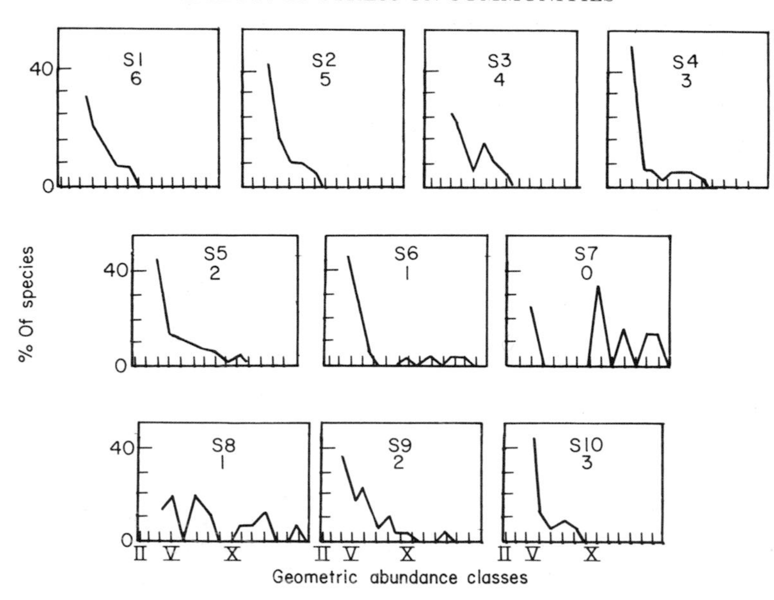

Figure 5. Plots of individuals per species along a transect across a sewage dumping ground in the Clyde estuary, Scotland. S1, S2 etc. sites; number below site shows distance in km from dump site. Individuals per species grouped into geometric classes, class I = 1 individual per species, class II = 2–3 individuals per species, class III = 4–7 individuals per species, class IV = 8–15 individuals per species etc. Data from Gray & Pearson, (1982).

eliminated and a few species become extremely common (the opportunist species). In most unstressed assemblages there are many rare species. My data for marine benthic assemblages in the Oslofjord show that 60–70% of the total species present occur at densities of 1 or 2–3 individuals per species per sampling unit, at any time-interval studied. There are no patches of high abundance. Yet rareness cannot be 'explained' in that it may result from specific habitat requirements, stochastic action of environmental factors on low density populations, random and low density of recruitment, patchy predation or competition etc. So it does not seem profitable to seek common patterns in the elimination of rare species. Instead the change in abundance of groups of moderately common species may be more profitable.

Gray & Pearson (1982) show how a group of species (usually 8–10) can be selected objectively from the total species list, (often 1–200 species in marine benthic assemblages). Table 2 shows a summary based on application of this technique to five different areas showing the moderately common benthic species that change in abundance in response to a variety of stressors. The suprising finding is that although the stressors vary from organic enrichment, to oil (a carbon source and therefore, an enrichment) and to TiO_2 waste there are a number of species that are common even to different geographical areas. There is no obvious pattern among the species in either feeding type nor common life-history traits, but it must be stressed that our knowledge of the general natural history of marine species is sorely lacking. Clearly a group of species has been identified that deserves further intensive study.

In the marine benthic examples shown above, the species that dominate under the most severe influence of stressors are opportunists. The polychaete *Capitella*

TABLE 2. Occurrence of selected moderately common species in stressed marine environments

	Area				
Species	Loch Eil (Pulp Mill)	Clyde (Sewage)	North Sea (Oil)	Oslofjord (Sewage)	Norway (TiO_2 Waste)
Lumbrinereis spp.		+		+	+
Pholoe minuta	+	+	+	+	+
Glycera spp.	+	+	+	+	+
Goniada maculata	+			+	+
Diplocirrus glaucus	+			+	
Anaitides spp.	+		+	+	
Chaetozone setosa		+		+	+

capitata occurs almost universally where there are large amounts of organic material in the sediment (Pearson & Rosenberg, 1978). Where there are large amounts of organic material there is usually low oxygen tension in the water, yet surprisingly *C. capitata* is not particularly tolerant of low oxygen tension (Reish, 1970). Nevertheless, *C. capitata* has many attributes of a typical opportunist in being small, having an extensive breeding season and producing between 10 000 and 14 000 eggs per female (Warren, 1976), having larvae that can be planktonic or benthic (Rasmussen, 1956) and having a 30–60 day life cycle (Reish & Barnard, 1960; Grassle & Grassle, 1976a). *Capitella capitata* is thus ideally suited to recolonizing freshly disturbed sediment. Indeed it was the first colonizer following an oil spill in Massachusetts (Sanders *et al.*, 1980) and following laying of a submarine pipeline (Eagle & Rees, 1973). Gray (1979) suggested that the reason why *C. capitata* dominates in organically enriched habitats is that there is continuous disturbance from settling particulate material which smothers most species. The surviving species are those which can recolonize the sediment quickly and *C. capitata* is a supreme example. Thus species that dominate under high organic enrichment in marine benthic sediments fit Grime's (1977) model of adaptive strategies of plants in that the primary effect is destruction of biomass by smothering and that in such cases an 'r' selected opportunist will be at an advantage.

In their review of effects of organic enrichment on marine organisms, Pearson & Rosenberg (1978) list as other common dominant species the polychaetes *Polydora ciliata* and *P. ligni*, *Heteromastus filiformis* and the oligochaete *Peloscolex, benendeni*. One of the traits of an opportunist (e.g. *C. capitata*) is the ability to have almost continuous breeding. In the Oslofjord there is a strong organic enrichment gradient from the city towards the outer fjord, with areas outside the critical sill at Drøbak being unaffected by organic matter pollution. Studies on the larvae of *P. ciliata* (Schram, 1970) show an interesting temporal adaptation to enrichment. Near Oslo, larvae are found for 12 months of the year, at Steilene (12 km from the city) for seven months and outside the sill at Drøbak only for four months, presumably correlated with the sedimentation rates of organic material and the recolonizing ability of the larvae.

In the Oslofjord there are six species of the genus *Polydora*, four of which co-occur at all sites and five co-occur at the two outermost unstressed sites (Schram, 1970). Recent studies on *Capitella capitata* (Grassle & Grassle, 1976b) show that from the Massachusetts coast there are not one but six genetically distinct species,

TABLE 3. Occurrence of sibling species in stressed marine habitats. *Heteromastus* and *Peloscolex* are not yet known to have species complexes

Genus	Stressor		
	Organic Enrichment	Hydrocarbon	TiO_2
Capitella	+	+	+
Heteromastus/* *Mediomastus*/ *Capitomastus*	+	+	+
Ophryotrocha	+	+	+
Polydora	+	+	
*Peloscolex**	+		

*Suspected species complex.

as shown by electrophoretic and breeding studies. The species differ additionally in egg size and larval length. Studies on *C. capitata* from the Oslofjord show that there are at least three different species from one locality.

Within 500 m of the Statfjord oil-platform (Olsgaard, unpublished observations) there are large numbers of the polychaete *Ophryotrocha* (which also occurs commonly under heavy organic enrichment) with many sibling species (Åkesson, 1973.) Thus one of the characteristics of marine benthic assemblages that are exposed to high stressor loads is the presence of species complexes. Table 3 shows the occurrence of known and suggested sibling species complexes found under high stressor load. All species shown are supposedly cosmopolitan, of small size, and colonize disturbed areas rapidly. *Heteromastus, Mediomastus,* and *Capitomastus* are all, like *Capitella,* within the family Capitellidae and are often misidentified or confused in monitoring surveys. The capitellid family in particular clearly responds to disturbance in a characteristic way. Whether or not *Peloscolex* is a complex of closely related sibling species remains to be seen. There are other known examples of opportunistic species that occur as species complexes—weeds, weevils (Suomalainen & Saura, 1973), aphids (Hille Ris Lambers, 1966) and brine shrimp (Barigozzi, 1974).

Why should there be species complexes in the most disturbed marine habitats? Grassle & Grassle (1977) suggest that in such groups the taxonomic unit that survives through evolutionary time may be a metaspecies where species are continuously being formed and becoming extinct. Templeton (1980, 1981) has reviewed the genetical mechanisms leading to speciation. The primary division is between divergence, where isolating barriers evolve in a genetically continuous fashion with selection leading to reproductive isolation, and transilience, where changes occur that lead to genetical discontinuities in the population. Templeton believes that adaptive divergence is the predominant divergent mode and occurs where a population is split into two sub-populations by an extrinsic barrier. Of the transilient modes, hybridization is thought to be the predominant mechanism although genetic transilience occurs in some groups.

Where there are high sedimentation rates, it is likely that isolation barriers are erected and local populations established. With a benthic larval development this gives possibilities for speciation by adaptive, clinal or habitat divergence. Yet there is little allelic similarity between the species of *Capitella* studied by Grassle & Grassle (1976b) which suggests that chromosomal events have been involved in speciation. The most likely mechanisms are by genetic transilience or founder flush. In genetic transilience a small founder population is established from an outcrossing, polymorphic and coadapted ancestral population. Alteration in the frequency of one or a few major alleles occurs and strong selective forces during the initial population expansion leads to a shift to a new adaptive peak. In the founder-flush model (Carson, 1968), during the population growth selection is thought to be weak, and strong selection occurs when the environment is saturated. Clearly more work needs to be done to establish the likely mechanism of speciation in *Capitella*.

In addition the 'species' *C. capitata* is reported to be cosmopolitan (Grassle & Grassle, 1976b). If the mechanism described above applies, one or more widely distributed species can be expected to provide the founder populations for the formation of locally distributed species.

Speciation within species complexes is not the only interesting aspect emerging from the marine data. There is no reason to suspect that the *Polydora* complex has speciated in response to high rates of disturbance as the species are easily identifiable and are found at many different localities. The presence of species in environments that are strongly influenced by stressors suggests that such species are tolerant of the conditions. Whether or not *Polydora* species show different tolerances of low oxygen and thereby illustrate Grime's tolerant adaptive strategy has not been studied.

Bradshaw and his co-workers (this volume) have in many elegant studies shown genetic adaptation to heavy metal tolerance in plant species. No similar work has been done on marine species. Possibly the temporal variation in larval occurrence in *P. ciliata* in Oslofjord, (described above) is a result of gene flow from the tolerant inner fjord population.

There is another interesting parallel between the effects of stressors on common and rare marine and terrestrial species obtained by comparing Hodgson's (1986a, b, c) studies on plants in the Sheffield area of Britain. Common plant species in the Sheffield area are those inhabiting sites of high fertility which are heavily disturbed and such plants have wide geographical ranges. Such species have a great capacity to exploit fertile habitats by prolific regeneration by seed. This description applies equally to *C. capitata*. Rare plant species in the Sheffield area are confined to infertile and relatively undisturbed habitats, again similar to habitats for rare species on the sea bed of Oslofjord.

REDUCTION IN SIZE

The mean size of organisms decreases along organic enrichment gradients (Pearson & Rosenberg, 1978). *Capitella* is much smaller than the large polychaetes that occur in undisturbed habitats. Similar trends are found in plants along gradients of irradiation or burning (Woodwell, 1967) with small-sized species occurring under the strongest influence of the stressor. When oil was added to a mesocosm with natural marine phytoplankton the small-sized organisms increased in abundance compared with the control (Parsons, Li & Waters, 1976). Similar results have been obtained by adding copper (Gamble, Davies & Steele, 1977) to a comparable system. The change to dominance by small species of phytoplankton is accompanied by replacement of the normally dominant large copepods in the zooplankton by smaller species (Gamble *et al.*, 1977).

Observations of long-term changes of plankton in the North Sea show that there has been a decline in large diatoms and increase in microflagellates and a decline in zooplankton biomass (Reid, 1975). This led Fisher (1976) to suggest that the North Sea ecosystem was responding to pollutants. Greve & Parsons (1977) took this argument further by comparing mesocosm and field data (notably from the North Sea) suggesting two alternative states for the marine food web. In the unstressed system large diatoms are the preferred food of large copepods which in turn provide food for young fish. In the stressed system small flagellates lead to small zooplankton species and ctenophores and medusae rather than fish. Observations that the frequency of occurrence of blooms of ctenophores and medusae has increased in the past few years in the North Sea has led to the suggestion that the stressed state was beginning to dominate.

The crucial question then is whether or not the changes observed in the North Sea are caused by man-made stressors or are natural processes. Dominance by small-sized phytoplankton species can be induced in lakes by altering the stability of the stratified layer in the summer period. Reynolds (1983) artifically perturbed the stratification and was able to maintain dominance of the system by a series of small-sized opportunist species whereas the natural sequence would have led to large-sized, slow growing, K-selected species dominating in summer. Similar processes occur in the sea (Harris, 1986). In the North Sea the abundance of diatoms declined between 1948 and 1970 and the growing period became shorter (Robinson, 1983). Diatoms need vertical turbulence to maintain the heavy cells in the surface layers. Up to 1969 the surface waters had become progressively warmer so that stratification occurred earlier and diatoms sedimented out of the system earlier, allowing the smaller r-selected species to dominate (Harris, 1986). Similar trends have been found in the Great Lakes (Mackarewicz & Baybutt, 1981). In the North Sea small copepods increased over the same period that the diatoms declined (Colebrook, 1978a, b). Cushing (1982) has shown convincingly that the changes in the North Sea are correlated with changed climatic patterns with a cooling trend since 1945 being the dominant pattern yet with warm years around 1950, 1960 and 1970 giving peaks of abundance over these years. Cushing (1984) has analysed in detail the correlations between climatology, biological events in the sea and recruitment to the haddock and cod stocks. There is little doubt that climate is the key controlling factor.

The Greve and Parsons model should, therefore, incorporate climate as an important stressor and cannot signify an effect of pollution-induced stress alone.

Finally Steele (1985) has compared the differences between temporal variability in terrestrial and marine ecosystems. He concludes that the important forcing physical factors will be different in the two systems. In the terrestrial system environmental variability is large at both short-and long-term, yet variance is constant per unit frequency up to periods of 50 years (0.02 cycles year^{-1}). Mechanisms will be selected therefore, which cope with short-term variability and minimize effects of longer-term variations. In marine systems environmental variability is of smaller amplitude over short-term periods, due to the large heat capacity of the ocean, but unlike the terrestrial system variance increases with period up to 1000–10 000 yrs which gives large amplitude changes at long time intervals and reflects the long period exchange rates between deep and surface waters. Marine organisms will therefore, show different responses to both short- and long-term variability than terrestrial organisms. The predominance of larval reproduction in the sea with the potential for dispersal and large scale gene flow is also radically different to terrestrial environments.

In conclusion, then, although there are similarities in responses to stressors in marine and terrestrial systems there are also fundamental differences. Separating out the effects on assemblages of stressors that act over relatively short-term periods such as pollution from climatic induced effects is a major problem for future research. Environmental variability associated with climatic variation probably acts as the ultimate stressor on both the ecological and evolutionary time scales.

REFERENCES

ADDISON, R., 1984. Hepatic mixed function oxidase (MFO) induction in fish as a possible biological monitoring system. In V. W. Cairns, P. V. Hodgson & J. O. Nriagu (Eds), *Contaminant Effects on Fisheries:* 51–60. Toronto: Wiley.

ADDISON, R. & EDWARDS, A. J., 1988. Hepatic microsomal monooxygenase activity in flounder (*Platichthys flesus*) from polluted in Langesundfjord and from mesocosms experimentally dosed with diesel oil and copper. In B. L. Bayne, K. R. Clarke & J. S. Gray, *Biological Effects of Pollutants. Marine Ecology Progress Series Special Issue, 46:* 51–54.

ÅAKESSON, B., 1973. Reproduction and larval morphology of five *Ophryotrocha* species (Polychaeta, Dorvilleidae). *Zoologica Scripta, 2:* 145–155.

BARIGOZZI, C., 1974. *Artemia:* a survey of its significance in genetic problems. *Evolutionary Biology, 7:* 221–262.

BAYNE, B. L., 1980. Physiological measurements of stress. In A. D. McIntyre & J. B. Pearce (Eds), *Biological Effects of Marine Pollutants and the Problems of Monitoring. Rapports et Proces–verbaux des Réunions de Conseil Permanent Internationale pour l'Exploration de la Mer, 179:* 56–61.

BAYNE, B. L. & WORRAL, C. M., 1980. Growth and production of mussels *Mytilus edulis* from two populations. *Marine Ecology Progress Series, 3:* 317–328.

BEUKEMA, J. J. (in press). An evaluation of Warwick's abundance/biomass comparison (ABC) method applied to macrozoobenthic communities living on tidal flats in the Dutch Wadden Sea. *Marine Ecology Progress Series.*

BROWN, D. A., BOWDEN, C. A., CHATEL, K. W. & PARSONS, T. R., 1977. The wildlife community of Iona Island, Jetty, Vancouver, B.C., and heavy-metal pollution effects. *Environmental Conservation, 4:* 213–216.

BROWN, D. A. & PARSONS, T. A., 1978. Relationships between cytoplasmic distributions of mercury and toxic effects to zooplankton and Chum salmon (*Onchorhynchus keta*) exposed to mercury in a controlled ecosystem. *Journal of the Fisheries Research Board of Canada, 35:* 880–884.

BUSBY, D. G., PEARCE, P. A. & GARRITY N. R., 1981. Brain cholinesterase response in songbirds exposed to experimental fenitrothion spraying in New Brunswick, Canada. *Bulletin of Environmental Contamination and Toxicology, 17:* 752–758.

CARSON, H. L., 1968. The population flush and its genetic consequences. In R. C. Lewontin (Ed.), *Population Biology and Evolution:* 123–137. New York: Syracuse University Press.

COLEBROOK, J. M., 1978a. Changes in the zooplankton of the North Sea, 1948 to 1973. *Rapports et Proces–verbaux des Réunions de Conseil Permanent Internationale pour l'Explorations de la Mer, 172:* 390–396.

COLEBROOK, J. M., 1978b. Continuous plankton records: zooplankton and environment, North-East Atlantic and North Sea, 1948–1975. *Oceanologica Acta, 1:* 9–23.

CUSHING D. H., 1982. *Climate and Fisheries.* London: Academic Press.

CUSHING D. H., 1984. The gadoid outburst in the North Sea. *Journal de Conseil Permanent Internationale pour l'Explorations de la Mer, 41:* 159–166.

EAGLE, R. A. & REES, E. I. S., 1973. Indicator species–a case for caution. *Marine Pollution Bulletin, 4:* 25.

FISHER, N. S., 1976. North Sea phytoplankton. *Nature, 259:* 160.

GAMBLE, J. C., DAVIES, J. M. & STEELE, J. H., 1977. Loch Ewe bag experiment, 1974. *Bulletin of Marine Science, 27:* 146–175.

GRASSLE, J. F. & GRASSLE, J. P., 1976a. Opportunistic life histories and genetic systems in marine benthic polychaetes. *Journal of Marine Research, 32:* 253–284.

GRASSLE, J. P. & GRASSLE, J. F., 1976b. Sibling species in the marine pollution indicator *Capitella* (Polychaeta). *Science, 192:* 567–569.

GRASSLE, J. F. & GRASSLE, J. P., 1977. Temporal adaptations in sibling species of *Capitella*. In B. C. Coull (Ed.), *Ecology of Marine Benthos.* Belle W. Baruch Library in Marine Science 6, S. Carolina. Columbia: University Press.

GRAY, J. S., 1979. Pollution-induced changes in populations. *Philosophical Transactions of the Royal Society, Series B, 286:* 545–561.

GRAY, J. S., 1982. Effects of pollutants on marine ecosystems. *Netherlands Journal of Sea Research, 16:* 424–443.

GRAY, J. S. & PEARSON, T. H., 1982. Objective selection of sensitive species indicative of pollution-induced changed in benthic communities. I. Comparative methodology. *Marine Ecology Progress Series 9:* 1121.

GREVE, W. & PARSONS, T. H., 1977. Photosynthesis and fish production: Hypothetical effects of climatic and pollution. *Helgolander wissenschaftliche Meeeresuntersuchungen, 30:* 666–672.

GRIME, J. P., 1977. Evidence for the existence of three primary strategies in plants and its relevance to ecological and evolutionary theory. *American Naturalist, 111:* 1169–1194.

HARRIS, G. P., 1986. *Phytoplankton Ecology: Structure, Function and Fluctuation.* London: Chapman and Hall.

HILLE RIS LAMBERS, D., 1966. Polymorphism in Aphididae. *Annual Review of Entomology, 11:* 47–78.

HODGSON, J. G., 1986a. Commonness and rarity in plants with special reference to the Sheffield flora. Part I: The identity, distribution and habitat characteristics of the common and rare species. *Biological Conservation, 36:* 199–252.

HODGSON, J. G., 1986b. Commonness and rarity in plants with special reference to the Sheffield flora. Part II: The relative importance of climate, soils and land use. *Biological Conservation, 36:* 253–274.

HODGSON, J. G., 1986c. Commonness and rarity in plants with special reference to the Sheffield flora. Part III: Taxonomic and evolutionary aspects. *Biological Conservation, 36:* 275–296.

LUDKE, J. L., HILL, E. F. & DIETER, D. P., 1975. Cholinesterase (ChE) response and related mortaility among birds fed ChE inhibitors. *Archives of Environmental Contamination Toxicology, 3:* 1–21.

MACKAREWICZ, J. C. & BAYBUTT, R. I., 1981. Long term (1927–1978) changes in the phytoplankton of Lake Michigan at Chicago. *Bulletin of the Torrey Botanical Club, 108:* 240–254.

McINTYRE, A. D. & PEARCE, J. B. (Eds), 1981. Biological effects of marine pollution and problems of monitoring. *Rapports et Proces–verbaux des Réunions de Conseil Permanent Internationale pour l'Exploration de la Mer, 179.*

McVIKAR, A. H., BRUNO, D. W. & FRASER, C. O., 1988. Fish disease in the North Sea in relation to sewage sludge dumping. *Marine Pollution Bulletin, 19:* 169–173.

MOLLER, H. & ANDERS, K., 1986. *Diseases and Parasites of Marine Fishes.* Moller: Kiel.

PATRICK, R., 1972. Benthic communities in streams. *Transactions of the Connecticut Academy of Arts and Sciences, Philadelphia, 259.*

PARSONS, T. R., LI, W. K. W. & WATERS, R., 1976. Some preliminary observations on the enhancement of phytoplankton growth by low levels of mineral hydrocarbons. *Hydrobiologia, 5:* 85–89.

PEARSON, T. H. & ROSENBERG, R. 1978. Macrobenthic succession in relation to organic enrichment and pollution of the marine environment. *Oceanography and Marine Biology Annual Review, 16:* 229–311.

RAPPORT, D. J., REGIER, H. A. & HUTCHINSON, T. C., 1985. Ecosystem behaviour under stress. *American Naturalist, 125:* 617–640.

RASMUSSEN, E., 1956. Faunistic and biological notes on marine invertebrates. III. The reproduction and larval development of some polychaetes from the Isefjord, with some faunistic notes. *Biologiske Meddelelser, 23:* 1–84.

REGIER, H. A. & HARTMAN, W. L., 1973. Lake Erie's fish community: 150 years of cultural stresses. *Science, 180:* 1248–1255.

REID, P. C., 1975. Large scale changes in North Sea phytoplankton. *Nature, 257:* 217–219.

REISH, D. J., 1970. The effects of varying concentrations of nutrients, chlorinity and dissolved oxygen on polychaetous annelids. *Water Research, 4:* 721–735.

REISH, D. J. & BARNARD, J. L., 1960. Field toxicity tests in marine waters utilizing the polychaetous annelid *Capitella capitata* (Fabricius). *Pacific Naturalist, 1:* 1–8.

REYNOLDS, C. S., 1983. A physiological interpretation of the dynamic responses of populations of a planktonic diatom to physical variability in the environment. *New Phytologist, 95:* 41–53.

ROBINSON, G. A., 1983. Continuous plankton records: phytoplankton in the North Sea 1958–1980, with special reference to 1980. *British Phycological Journal, 18:* 131–139.

ROUGHGARDEN, J., IWASA, Y. & BAXTER, C. 1985. Demographic theory for an open marine population with space-limited recruitment. *Ecology, 66:* 54–67.

SANDERS, H. L., GRASSLE, J. F., HAMPSON, G. R., MORSE, L. S., GARNER-PRICE, S. & JONES, C. C., 1980. Anatomy of an oil-spill: long term effects from the grounding of the barge *Florida* off West Falmouth, Massachusetts. *Journal of Marine Research, 38:* 265–381.

SCHINDLER, D. W., 1987. Detecting ecosystem responses to anthropogenic stress. *Canadian Journal of Fisheries and Aquatic Science, 44:* 6–25.

SCHRAM, T., 1970. Studies on the meroplankton in the inner Oslofjord. II. Regional differences and seasonal changes in the specific distribution of larvae. *Nytt Magasin for Zoologi, 18:* 1–21.

SELYE, H., 1973 The evolution of the stress concept. *American Science, 61:* 692–699.

STEELE, J. H., 1985. A comparison of terrestrial and marine ecological systems, *Nature, 313:* 355–358.

STEGEMAN, J. J., 1980. Mixed-function oxidase studies in monitoring for effects of organic pollution. In A. D. McIntyre & J. B. Pearce (Eds), *Biological Effects of Marine Pollution and the Problems of Monitoring. Rapports et Proces–verbaux des Réunions de Conseil Permanent Internationale pour l'Exploration de la Mer, 179:* 33–38.

STEGEMAN, J. J., WOODIN, B. R. & GOKSØYR, A. 1988. Cytochrome P-450 induction in flounder. In B. L. Bayne, K. R. Clarke & J. S. Gray, *Biological Effects of Pollutants. Marine Ecology Progress Series Special Volume, 46:* 55–60.

SUOMALAINEN, E. & SAURA, A., 1973. Genetic polymorphism and evolution in parthenogenetic animals. I. Polypoid Curculionidae. *Genetics, 74:* 489–508.

TEMPLETON, A. 1980. Modes of speciation and inferences based on genetic distances. *Evolution, 34:* 719–729.

TEMPLETON, A., 1981. Mechanisms of speciation—A population genetic approach. *Annual Review of Ecology and Systematics. 12:* 23–48.

VERMA, S. R., TONK, I. P., GUPTA, A. K. & DALELA, R. C., 1981. *In vivo* enzymatic alterations of certain tissues of *Sacchobranchus fossilis* following exposure to four toxic substances. *Environmental Pollution, Series A. Ecological Biology, 26:* 121–127.

WARWICK, R. M., 1986. A new method for detecting pollution effects on marine macrobenthic communities. *Marine Biology, 92:* 557–562.

WARWICK, R. M., PEARSON, T. H. & RUSWAHYUNI, 1987. Detection of pollution effects on marine macrobenthos: further evaluation of the species/abundance method. *Marine Biology, 95:* 193–200.

WARREN, L. M., 1976. A population study of the polychaete *Capitella capitata* at Plymouth. *Marine Biology, 38:* 209–216.

WIDDOWS, J. & JOHNSON, D., 1988. Physiological energetics of molluscs. In B. L. Bayne, K. R. Clarke & J. S. Gray (Eds), *Biological Effects of Pollutants. Marine Ecology Progress Series Special Volume, 46:* 113–121.

WILLIAMS, W. T., 1980. Air pollution disease in the Californian forests: a base-line for smog disease on ponderosa and Jeffrey pines in the Sequoia and Los Padres national forests, California. *Environmental Science and Technology, 14:* 179–182.

WOODWELL, G. W., 1967. Radiation and the patterns of nature. *Science, 156:* 461–470.

WULFF, F., 1979. The effects of sampling frequency on estimates of the annual pelagic primary production in the Baltic. In H. Hytteborn (Ed.), *The Use of Ecological Variables in Environmental Monitoring*. The Swedish Environment Protection Board, Report PM 1151: 147–150.

Biological Journal of the Linnean Society (1989), *37:* 33–49. With 2 figures

Symptoms of pathology in the Gulf of Bothnia (Baltic Sea): ecosystem response to stress from human activity

DAVID J. RAPPORT

Statistics Canada, Ottawa, Ontario, K1A OT6, Canada

An extensive review of the data now accumulated on the response of the Gulf of Bothnia ecosystem to stress from human activity provides abundant evidence for incipient ecosystem pathology. Signs of 'ecosystem distress' appear at local, coastal and basin-wide scales. Symptoms include early signs of eutrophication in local and coastal waters, formation of local abiotic zones, reduction in species diversity, reduction in genetic diversity (particularly in salmonids), reduced size of biota, increased dominance by opportunistic species, increased disease prevalence and bio-accumulation of toxic substances (e.g. PCBs, DDT, heavy metals). Pathology may propagate between local, coastal and basin-wide regions by several different pathways. Despite reductions in loadings of some toxic substances in recent years (e.g. PCBs, DDT), stress from human activities on the Gulf of Bothnia continues to impact the ecosystem at all spatial scales.

KEY WORDS:—Baltic Sea – Gulf of Bothnia – ecosystem degradation – ecosystem behaviour under stress – state of environment – ecosystem pathology.

CONTENTS

INTRODUCTION

The observation that human activity can profoundly affect the state of nature is a very ancient one. It dates back, at least, to the writings of Plato who commented on the fact that cultivation altered (adversely) the local climate.

This work is dedicated to the late Svante Oden, whose imaginative approaches and discoveries of mechanisms of environmental pathologies have set the highest standards for work in the area of stress on ecosystems.

0024–4066/89/050033 + 17 $03.00/0

Today, while local impacts of human activity are of considerable importance, more and more concerns are shifting to the regional, landscape and global levels (e.g. Clark & Munn, 1986).

A comparative review of case histories of stress on large ecosystems (e.g. Central European forests, Laurentian Great Lakes, Baltic Sea) suggests that there are common symptoms of pathology which I have previously referred to as the 'ecosystem distress syndrome' (Rapport, Regier & Hutchinson, 1985). Table 1 compares these findings with two other studies: a theoretical study of the impact of stress on ecosystem development (Odum, 1985) and a study of the impact of economic and cultural change on the landscape (Godron & Forman, 1983).

While the specific indicators of ecosystem response to stress differ to some extent among these studies, there are broad similarities in the following aspects: (1) there appears to be a reduction in the efficiency with which ecosystems process energy (reflected variously as a decline in community respiration, primary production, or net landscape production); (2) there is an increase in the horizontal flow rates of nutrients; that is terrestrial systems lose nutrients and aquatic systems accumulate them; (3) changes in community structure appear to favour biota that are short-lived, smaller, exotic, and have high reproductive rates. Such features often characterize 'weedy' or 'pest' species; (4) ecosystem

TABLE 1. Response to stress at ecosystem and landscape levels, as identified in three studies —, Decrease; +, Increase; ?, not specified in source study; *, symptom present

System property	Ecosystem (Odum, 1985)	Ecosystem (Rapport *et al.*, 1985)	Landscape (Godron & Forman, 1983)
Energetics			
Community respiration	—	?	?
Primary production	?	—	?
Net landscape production	?	?	—
Nutrient flow			
Horizontal transport	+	+	+
Community structure			
Species diversity	—	—	—
r-selected species	+	+	+
Short-lived species	+	+	?
Smaller biota	+	+	?
Food chain length	—	?	?
Exotic species	?	+	?
System features			
Openness	+	+	+
Succession reversal	*	*	?
Metastability	?	?	—
Negative interaction	+	?	?
Disease incidence	?	+	?
Mutualism	—	?	?
Population self-regulation	?	—	?
Resource use efficiency	—	?	?
Boundary distinctiveness	?	?	+
Boundary linearity	?	?	+

From Rapport & Regier, 1989.

development appears to reverse in direction, that is to 'retrogress' to resemble earlier stages in which self-regulatory functions are less developed, species diversity is reduced, etc.

Taking the existence of a distress syndrome at the ecosystem level as a working hypothesis, I examined data on conditions and trends in the Gulf of Bothnia (Baltic Sea). Two questions were posed: (1) what evidence might be found for the existence of a distress syndrome at the ecosystem level and (2) what is the temporal/spatial history of the development of the distress syndrome?

STRESS ON THE GULF OF BOTHNIA

The Gulf of Bothnia (Fig. 1) is situated in the northern-most part of the Baltic Sea (59–66°N,17–26°E). It is the least saline area of the Baltic, owing to the large volume of freshwater run-off and its relative isolation from the southern Baltic. It is a young ecosystem, as is the entire Baltic Sea, having been created

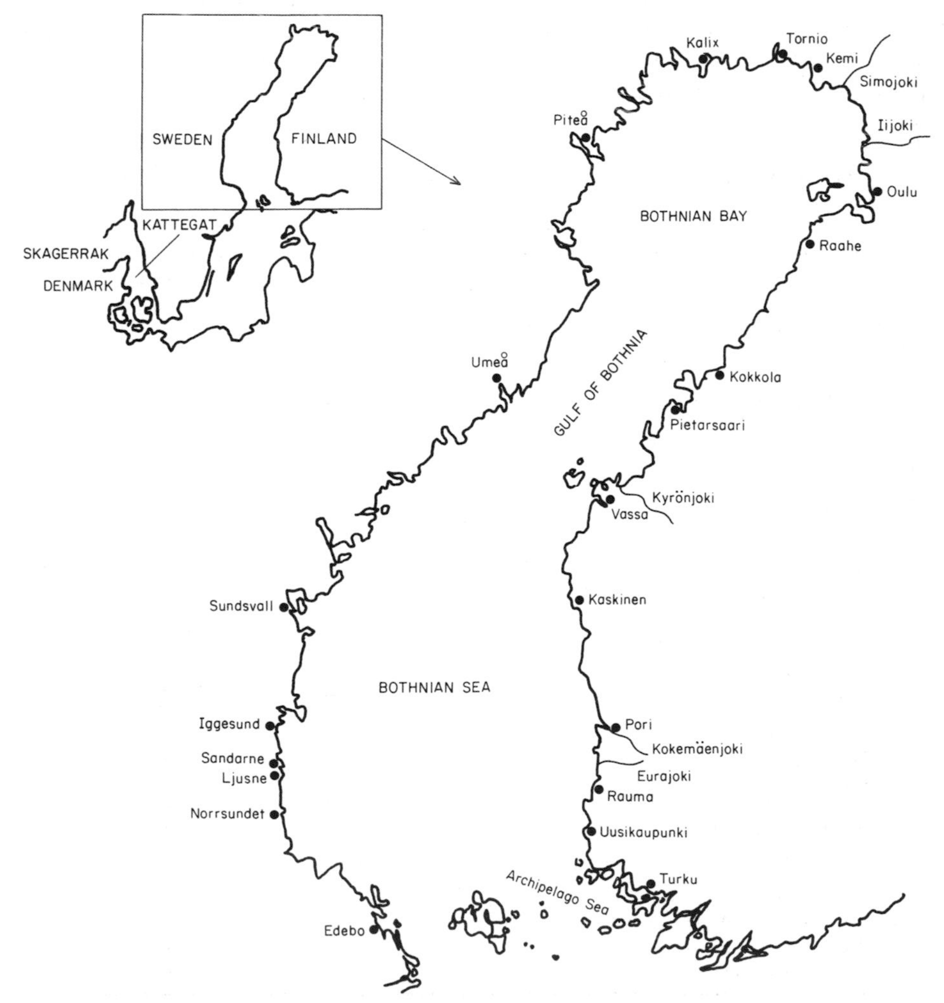

Figure 1. The Gulf of Bothnia with reference to study sites mentioned in the text.

c. 10 000 years ago with the melting of the ice cap from the last ice age (Voipio, 1981).

Winters are long and cold, with ice formation beginning in the open Bothnian Bay in early January, spreading to coastal zones of the Bothnian Sea by mid-January, and extending to the southern part of the Gulf by mid-February (Palosuo, 1966). The Bothnian Bay is ice-covered 100–150 days a year. In five winters out of ten, the Bothnian Sea is also ice-covered.

The Baltic is a shallow sea, the deepest parts being small in area and less than 500 m. Much of the Gulf is less than 50 m in depth, forming a particularly shallow zone along the most northern part and along the entire Finnish coast. Owing to the complex topography of the coastline with its innumerable bays, estuaries and islands, the coastline is many times its *c.* 750 km rectilinear measure. The Finnish coastline alone (but including the Gulf of Finland) is *c.* 39 000 km (Pitkänen *et al.*, 1987). On this basis, the Finnish and Swedish coastlines of the Gulf of Bothnia exceed 50 000 km.

There are few population centres of any significant size along the coast; only one, the Turku-Naantali area, has a population greater than 100 000 (Rikkinen, 1980). Heavy industries, particularly on the Swedish side are concentrated by the sea (Bruneau, 1980); these include steel and metal works, pulp and paper and chemical plants, many of which have had environmental impacts over large regions.

The character of the Gulf is dominated by three factors. First is the relatively young age of the Baltic Sea, which in itself accounts for the fact that biotic diversity is relatively low. Second, the large temperature gradient from the nothern part of the Bothnian Bay to the Archipelago Sea, coupled with the substantial gradient in hours of sunlight, accounts for large differences in primary and secondary productivity. Finally the low salinity (which ranges from less than three parts per thousand in the north to seven parts per thousand in the south) limits the distribution of marine species and permits the retention of a substantial freshwater community in coastal areas and in much of the Bothnian Bay.

The low salinity regime coupled with cold temperatures and long dark winters have permitted the development of biotic communities which in many ways have the characteristics of stressed ecosystems—for example, low species diversity, low productivity, inefficient nutrient exchange (Voipio, 1981). For example, of *c.* 1500 macroscopic species found off the Norwegian coast, only 145 occur in the southern Baltic Proper and no more than 52 in the southern Bothnian Sea (Hällfors *et al.*, 1981).

These features make it more difficult to detect symptoms of anthropogenic stress, for if they are present they are superimposed on an already stressed environment. Further, natural fluctuations in conditions owing to significant seasonal changes and interannual variability make early detection of departures from 'norms' all the more difficult and uncertain.

Stress on the various drainage basins of the Gulf of Bothnia appears to have intensified considerably over the past 2–3 decades. A typical pattern for the northern Finnish coast (Bothnian Bay) is shown by changes in the Kyrönjoki estuary (Fig. 2A, B). Except for large-scale drainage of forest and peat lands, most of these stresses were evident, in some form or other, by the beginning of the 19th century. Intensification of activity over the past quarter century (reflecting

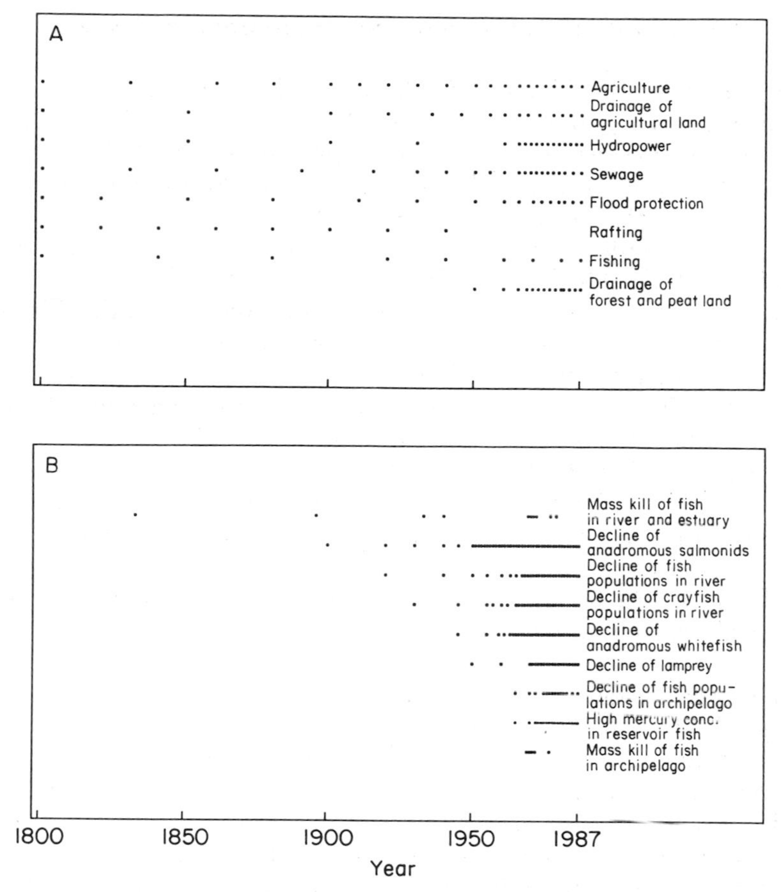

Figure 2. A. The development of human stresses on the Kyrönjoki ecosystem. The density of the dots corresponds to the intensity of particular stresses. Sparse dots indicate presence of an activity without large scale effects on the system. Dense dots indicate activity with stress impacts over large areas. (From Hildén & Rapport, 1989). B. Temporal sequence of signs of pathology in response to stress (Fig. 2A) in the Kyrönjoki system. Single dots refer to beginning or isolated occurrences. Dense dots indicate that the symptom has become a regular feature of the system. (From Hildén & Rapport, 1989).

both an expansion of settlements and the application of more invasive technologies) has greatly increased the pressures on the Gulf of Bothnia environment.

A 'MEDICAL' HISTORY

An overview of symptoms of incipient pathology in the Gulf of Bothnia is shown for 'local', 'coastal' and 'basin' regions. These classifications are for the sake of convenience and do not represent a rigorously-drawn system. Generally, 'local' refers to recipients of industrial sites or urban centres; for example, the straits off Turku are the recipient of effluents from the Finnish towns of Turku, Raisio and Naantali. Areas designated as 'coastal' are beyond the reach of protected waters; that is, to the seaward side of archipelagos, bays, estuaries. 'Basin' designates 'offshore' waters of the two major divisions of the Gulf of Bothnia—the Bothnian Bay in the north and the Bothnian Sea in the south.

TABLE 2. Symptoms of pathology in the gulf of Bothnia. ◆definite; ■, likely; Θ, perhaps; —, no data available; ▲, increase; ▼, decrease; *, change to favour opportunistic species; **, dominant influence from change in seal populations

Symptom	Change	Area		
		Local	Coastal	Basin
Nutrients	▲	◆	◆	◆
Productivity	▲	◆	■	Θ
Abiotic zones		◆	—	—
Species diversity	▼	◆	—	—
Genetic diversity	▼	◆	■	◆
Size distribution	▼	◆	■	■**
Disease prevalence	▲	■	—	■**
Biotic composition	*	◆	■	—

From Hildén & Rapport, 1989.

Nutrients

Evidence that nutrient concentrations are elevated locally, coastally and basin-wide derives from a variety of studies. Rivers are a major source of loadings, draining extensive agricultural and forest lands which, in addition to natural sources of nutrients, also contain significant input from agriculture, industry and urban areas (Larsson, Elmgren & Wulff, 1985). According to a survey of the Committee on the Gulf of Bothnia (Enckell-Sarkola *et al.*, 1984) the total load of phosphorus in the Gulf is *c.* 8000 tonnes per annum, of which *c.* 25% comes from the atmosphere. Of the 6000 tonnes comming from the drainage basin, approximately 20% is attributable to municipal and industrial effluents discharged directly into the sea. These discharges were greatest during the 1960s, but declined sharply in the early 1970s owing to environmental protection measures adopted by the pulp and paper industry and improved municipal sewage treatment.

The impact of industrial and municipal loadings on nutrient concentrations in bays and inlets can be substantial. For example, surface winter concentrations of total phosphorus averaged 11–13 mg m^{-3} in the Bothnian Sea, while registering over 100 mg m^{-3} in the heavily loaded inner Bay of Uusikaupunki (Pitkänen *et al.*, 1987).

In coastal waters, loads from the mainland generally dominate the areal distribution of nutrients. Further, the complex topography of coastal waters tends to reduce exchange with the offshore areas (Pitkänen, Tulkki & Kangas, 1985). Thus nutrient concentrations are elevated in coastal waters off estuaries, harbours, and in archipelagoes. Exceptions are the coastal waters of the central Bothnian Bay and the central Bothnian Sea on the Finnish coast, which are characterized by good mixing conditions and relatively low loadings (Pitkänen, 1986).

To address the impact of human activities on coastal nutrient levels, it is necessary to compare selected river basins in the same regions which are known to be relatively clean (none nowadays is absolutely pristine) with those that are known to be highly eutrophicated as a result of agricultural activities, industry and human settlements. Examining the data provided by Pitkänen *et al.*, (1985) on the variation and trends in nutrient concentrations in the waters around Finland, I compared values for the Eurajoki (a relatively clean water river compared with neighbouring rivers) with the polluted Kokemäenjoki. Both rivers discharge to the Bothnian Sea. The coastal region in the vicinity of the Kokemäenjoki has total phosphorus concentrations in the range of 30–35 mg m^{-3}, while near the Eurajoki phosphorus concentrations are in the range of 12–20 mg m^{-3}. Similarly, comparing river systems that discharge to the Bothnian Bay near the polluted Kyrönjoki, total phosphorus concentrations in surface water are in the range of 40–60 mg m^{-3}, while coastal waters near the relatively clean waters of the Iijoki and the Simojoki have values in the range of 15–20 mg m^{-3}.

A similar, although more dramatic, picture holds for nitrogen. Rather high coastal values in the north are due largely to the naturally humic rivers. However, in the Kyrönjoki area, dredging and agricultural drainage have greatly increased the levels—and they are among the highest given for the entire Finnish coast, being in the range of 600–1400 mg m^{-3} in winter surface waters. High values for the southern coast reflect both the natural nutrient-rich soils coupled with run-off from agricultural development in the region.

The situation is less clear when it comes to offshore waters. While there is evidence for elevated nutrient concentrations in the Baltic Proper (Larsson, Elmgren & Wulff, 1985), data for the Gulf of Bothnia are less convincing. However, trends derived from simple regressions on data from 'representative' stations are suggestive that nutrient concentrations have risen in comparison with the 1960s (Pitkänen *et al.*, 1987: 66, table 12).

Productivity

A rise in the nutrient supply is invariably associated with increases in productivity (Rapport *et al.*, 1985), unless there are special mitigating circumstances, such as the action of toxic substances which has tended to offset to some extent the effect of organic enrichment in areas such as Pietarsaari, Kaskinen, Pori and Rauma (Pitkanen *et al.*, 1987).

The clearest evidence for eutrophication is for local areas. Eutrophic conditions are found in many of the bays and estuaries on both coasts. For example, along the Finnish coast some 13 areas were designated as 'loaded areas' (Pitkänen *at al.*, 1987); that is, areas in which loadings have caused damage to the receiving local and coastal ecosystems. These areas are: Tornio, Kemi, Oulu, Raahe, Kokkola, Pietarsaari, the Kyrönjoki estuary, Vassa, Kaskinen, Pori, Rauma, Uusikaupunki and the Turku region (including the Archipelago Sea). In many of these, mean summer values for primary production capacity (PPC) exceeded 100 mg C m^{-3} d^{-1}. In a few cases, PPC exceeded 300 mg C m^{-3} d^{-1} (e.g. Oulu, Raahe). 'Non-loaded' coastal areas have summer mean PPC values of 67 mg C m^{-3} d^{-1} for the Bothnian Sea and 39 mg C m^{-3} d^{-1} for the Bothnian Bay. In some of the Finnish coastal areas, for example, the middle archipelago off Turku, for which there exists a long monitoring series, the data suggest an increase in primary productivity, changing the character of the area from a once oligotrophic system to one now considered slightly eutrophicated.

Cumulative effects of local eutrophication coupled with the more diffuse land run-off, particularly in agricultural areas, has resulted in the spread of eutrophication along coastal areas. Evidence for coastal eutrophication comes from the direct monitoring of concentrations of chlorophyll *a* (Pitkänen *et al.*, 1987; e.g. see above), surveys of Cladophora (Makinen & Aulio, 1987) and a comparison of historical and recent surveys of macrozoobenthos along the Swedish coast of the Bothnian Sea (Cederwall & Leonardsson, 1984).

Comparing the mean macrofaunal biomass of stations visited by Hessle (1924) with data collected in the 1960s and in the early 1980s at the same stations using similar techniques, Cederwall & Leonardsson (1984) found that at all depths (the range varied from less than 10 m to just over 100) for the central Bothian Sea the macrozoobenthos biomass increased in both periods. At the shallower depths, the increase was in the range of 7–8-fold over the first period of 40 years, a change which is unlikely to be totally accounted for by cyclical fluctuations in macrozoobenthos, which are in the range of 3–4-fold (Andersin, 1987).

A similar comparison for stations in the northern Bothnian Sea yielded less clear results, although here, too, most stations sampled in 1984 had greater macrofaunal biomasses than in 1924, and several stations (in this case at intermediate depths) had considerably greater biomass (Cederwall, personal communication).

Offshore waters of the Gulf of Bothnia are still considered oligotrophic. Mean chlorophyll *a* concentrations during July–September in the open sea were below 2 mg m^{-3} for all but one station (an anomaly, for it is located in the northern Bothnian Bay) for the summers 1979–83 (Pitkänen *et al.*, 1985). Given the large interannual variability in the measures of primary productivity and primary biomass, the best data to assess changes in production in the open sea are surveys of macrozoobenthic biomass, particularly the long series on the deep macrozoobenthos, collected by the Finnish Marine Institute (Andersin, 1987). These data suggest that despite great annual variation, the macrozoobenthos showed a gradual increase in biomass over the period 1965–83 in the dominant species in both the Bothnian Sea and Bothnian Bay. At many stations, the highest abundance and biomass values since the first sampling in the middle of the 1960s have been reported in the 1980s. Andersin (personal communication) cautions that these results are preliminary in that the data are unadjusted for seasonal differences in sampling and regional differences between stations. Moreover, the wet weight determinations have been done by different workers over the 20-year-period and as results can be sensitive to slight differences in procedure; this also adds to uncertainties in interpreting the data.

Abiotic zones

While the existence of abiotic zones might, in and of itself, suggest incipient pathology, such phenomena exist as a very natural condition in certain bodies of water where there is little mixing: for example, such regions are a prominent feature of the shallow central basin of Lake Erie (Laurentian Great Lakes) (Barica, 1982) and the Baltic Proper (Dybern & Fonselius, 1981). In these cases the onset of ecosystem pathology may be detected as an enlargement of the size of the naturally abiotic region.

For the Gulf of Bothnia there are few if any naturally occurring abiotic zones. Yet the evidence is convincing that many such zones have been created locally by the excessive nutrient and organic loadings of local industries and municipalities. Cederwall (1987), in his survey of sites along the Swedish Coast, found a number of areas in proximity to pulp mills (e.g. Kalix, Piteå, Umeå, Sundsvall, Iggesund, Bay of Edebo, Ljusne Bay) where macrofaunal benthos was absent. While these extend only a few square km, they are symptomatic of severe local pathology. Leppäkoski (1975) documents similar phenomena for local regions along the Finnish coast. No such zones (with one possible exception—Andersin *et al.*, 1987) have been detected in the coastal and offshore areas of the Gulf of Bothnia.

Species diversity

The only evidence for reductions in species diversity in the Gulf of Bothnia is at the local level. Here there are well documented losses of both macrozoobenthos and fish stocks owing to organic enrichment, the release of toxic substances, acidification of the drainage basin and other sources.

Leppäkoski's detailed studies (Leppäkoski, 1975) on the impacts of organic enrichment on estuarine soft-bottom communities in the straits off Turku document losses in macrozoobenthos. He identified five distinct variants of the original community along a gradient ranging from 'pollution tolerant' species to 'clean water species'; that is, species absent in the polluted areas. The community variants closest to polluted zones contained far less species than the communities nearer the 'clean water' end of the spectrum. Cederwall (1987) obtained similar results for local recipients of pulp and paper effluents along the Swedish coast.

Fish associations are also impoverished by pollution. For example, at Pori, a titanium dioxide industry discharges metals in solution, the most important being iron, titanium, aluminium and manganese. Fish kills have been reported within 4–6 km of the outlet, probably mainly due to the high concentrations of iron in the effluent and the low pH. Local catches of whitefish and herring have been reduced (Parmanne *et al.*, 1986). In the Kyrönjoki river estuary, the spring run-off has become more acidic due to drainage and agricultural practices which have served to mobilize the naturally acidic soils. While there have been occasional reports of dying fish for over a century, more frequent mass fish kills were reported since the late 1970s. Burbot and whitefish appear to have been lost to the estuary (Hildén & Rapport, 1989).

Similarly, off the Swedish coast at Norrsundet (Bothnian Sea), in the vicinity of a pulp mill producing bleached kraft pulp, abnormally low catches of fish were reported close to the outlet (Neuman & Karås, 1988). Compared with catches at Sandarne (the site of a mill producing unbleached pulp) both the abundance of fish and species diversity at Norrsundet was lower (six species at Norrsundet, ten at Sandarne). Similar impacts were demonstrated for other mill sites along the Swedish coast (Hansson, 1987).

There have been a number of introduced species to the Baltic which raises, although artificially, the species diversity. Leppäkoski (1984) provides a list of introduced species to the Baltic Sea. About one-third of them originate from North America and are mainly restricted to littoral or shallow sublittoral subsystems of the Southern Baltic region. A few species have become established in the Gulf of Bothnia (e.g. *Elodea canadensis, Mya arenaria, Branta canadensis*).

From the point of view of a conservationist, such introductions may be regarded as contamination of the biota by foreign elements. Whether this has contributed to the development of ecosystem pathology remains an open question. In the Great Lakes the facilitated introduction of the Sea Lamprey has certainly had a deleterious effect (Regier & Hartman, 1973).

Genetic diversity

Thus far in the Gulf of Bothnia, while species have not disappeared on a coastal or basin-wide scale, some have come perilously close (e.g. the ringed and grey seal populations). Genetic diversity, however, has declined; for example, the demise of the wild salmon and brown trout stocks. This is mainly a consequence of physical restructuring within spawning river systems (e.g. construction of river impoundments, channels, reservoirs, hydro-engineering activities) which are destructive to critical habitats. Of the 18 indigenous Atlantic salmon stocks spawning in rivers that flow into the Baltic, only two naturally spawning populations remain. Out of a total of 47 original stocks of sea trout, only five remain (Westman & Kallio, 1986).

Size distribution

Systematic changes in the size distribution of organisms to favour the smaller species often characterize the surviving communities in stressed environments (Woodwell, 1967). While definitive studies on such changes for the Gulf of Bothnia biota appear lacking, anecdotal evidence is readily found. For example, with respect to the macrozoobenthos it would appear that 'clean water' species (i.e. dominated by the amphipod, *Pontoporeia affinis* and the bivalve, *Macoma balthica*) are decidedly larger (on a volume basis) than 'dirty water species' (i.e. oligochaetes). Typically, with organic enrichment, the order of disappearance of macarozoobenthos is: (1) *Pontoporeia affinis;* (2) *Mesidothea entomon;* (3) *Harmothoe sarsi;* (4) *Macoma balthica.* These species which are the first to go are all the large to mid-size macrozoobenthos (S. Hansson, personal communication). There is also evidence that, under stressed conditions, the size composition within benthic species (e.g. *Macoma baltica*) shifts towards younger and smaller individuals (Skog & Varmo, 1980).

On a basin-wide scale, the top predators (birds of prey, baltic ringed seal and grey seal) in the Baltic have been much reduced in number (Helle, 1986). This observation also supports the hypothesis that stress results in reductions in the size spectrum of surviving orgnisms. The decline in seals alone would have this effect as they constitute the largest organisms in the Gulf.

Disease prevalence

Observations on diseases in Baltic fish (e.g. deformities such as occur in fourhorn sculpin; skin tumours such as occur in northern pike; fin rot such as occur in perch) suggest a possible correlation between disease prevalence and stress. For example, there are reports of eyeless Baltic herring in the vicinity of the titanium plant at Pori (Parmanne *et al.*, 1986), outbreaks of fin erosion in perch following an oil spill in the northern Bothnian Bay (Urho & Hudd, submitted), and spinal deformities in fourhorn sculpin off the Finnish coast of the Bothnian Bay associated with significant concentrations of heavy metals (Bengtsson & Miettinen, 1987).

Yet the hypothesis remains speculative for several reasons. First, some diseases, such as skin tumours in northern pike, are rare or absent even in highly polluted waters and it is suggested that physiological stresses rather than pollution might be a predisposing factor (Thompson, 1982). Secondly, some fish diseases associated with infections, such as fin erosion, appear correlated with polluted waters. However these diseases are also found in clean waters and in many instances we lack the necessary 'base-line' data to determine with any degree of certainty whether disease prevalence is elevated in polluted waters. Thirdly, there is the complication of unknown sampling biases introduced by possible changes in swimming behaviour of diseased fish (e.g. are diseased fish impaired in their swimming behaviour thereby becoming more easily caught in sampling gear?).

The picture is markedly different when it comes to reproductive failures in Baltic seals, suggesting increased disease prevalence in this important basin-wide component of the Baltic ecosystem. Pathological uterine occlusions (the frequency of which was as high as 60% among mature ringed seals in 1970–79; Helle, 1986) in both ringed seal (*Phoca hispida*) and grey seal (*Halichoerus grypus*) have been linked to high body burdens of PCB and DDT compounds (Helle, Olsson & Jensen, 1976; Bergman & Olsson, 1986). Helle (1986) suggests that the reproductive failures attributable to this disease have resulted in a decrease, by one-third, in the size of the ringed seal population in the Gulf of Bothnia over the period 1975–84.

Biotic composition

Most marked changes in biotic communities have occurred at local and coastal levels. How well do the observations support the hypothesis that such changes favour the opportunistic species, that is, the short-lived, high-fecundity variety?

Leppäkoski's work on the macrozoobenthos (Leppäkoski, 1975) again provides a rich data base upon which to examine this hypothesis. The restructuring of the initial macrozoobenthic community of the inner archipelago waters (Finland) can be viewed in several ways. First, pollution-indicator species dominate in the most impacted waters. These species, comprised of midge larvae and oligochaetes, have both a shorter life cycle and are smaller (by volume) compared to other members of the benthos assemblage. Secondly, there appears to be a displacement (seaward) of the marine and brackish-water species leaving an increased dominance of fresh water species inshore (Leppäkoski, 1979).

Systematic changes in the structure of fish communities at both local and coastal levels have also been observed. In the acidified waters of the Kyronjoki estuary, it is the smaller species (percids spp.) that survive the spring episodes of acidic stress, while the larger species (anadronomous salmonid spp.) are absent (Hildén & Rapport, 1989). In the vicinity of Oulu (Bothnian Bay), where the primary stress was organic enrichment, vendace, a coastal salmonid, declined and the community shifted towards the more tolerant smelt inshore, and the more tolerant herring farther out (Hildén, Lehtonen & Böhling, 1984). Vendace appear more sensitive to eutrophication since their eggs overwinter and thus are subjected to higher mortality in situations in which there is increased sedimentation.

Systematic shifts in the composition of fishes found in the vicinity of pulp mills have been shown for nine recipients (Karlsborg, Piteå, Husum, Ornskoldsvik, Kopmanholmen, Ostrand, Ortviken, Sandarne, Norrsundet) along the Swedish Coast (Neuman & Karŝ, 1988; Hansson, 1987). Close to the outlet, abundance of all species was very low. Farther out, fish biomass was elevated in a zone characterized by dense populations of ruffe and cyprinids, mainly roach. Other important species in the coastal community, such as sand-goby, occurred in below normal densities. The relationship of these changes to the emissions from pulp mills are very complex, as the mills emit substances which are eutrophicating, colouring, toxic and repelling. Such stresses, broadly speaking, are similar to those of eutrophication with the addition of repelling and toxic substances. Similar shifts in community structure are reported for sites along the Finnish coast: that is, increasing nutrient levels appear to have caused enhancement of coastal cyprinids and ruffe (Hildén, Hudd & Lehtonen, 1982; Hildén *et al.*, 1988). In the Archipelago Sea, pike-perch also appears to have benefitted from a moderate euthrophication (Hildén *et al.*, 1982).

Bioaccumulation of contaminants

One of the most easily documented signs of ecosystem pathology is the circulation and bioaccumulation of man-made persistent organic compounds. Any number of examples may be given: for example, concentrations of mineral oil in coastal waters near industrial sites; concentrations of resin acids and total chlorinated phenolics in pulp mill effluents; concentrations of PCB and total DDT in extractable fat of fish (e.g. Baltic herring and seals) etc. These substances, which do not occur naturally in ecosystems, are *prima facie* evidence of stress imposed by human activity as they have been shown to have quite deleterious consequences at physiological and population levels (Olsson, 1987; Olsson & Reutergardh, 1986; Pitkånen *et al.*, 1987). In healthy ecosystems, such substances might be sequestered in sediments or taken up by macrophytes or otherwise detoxified by various natural processes (Neuhold & Ruggerio, 1975). It is in the more stressed ecosystems that, owing to both higher loadings and some impairment in the mechanisms of detoxification, these substances more freely circulate.

The ecotoxicological effects of these and other toxic substances (e.g. heavy metals) have only been extensively investigated for organisms at the top of the food chain: for example, seals, white-tailed sea eagles, guillemot (Helle, 1986; Helander, Olsson & Reutergårdh, 1982; Olsson & Reutergårdh, 1986). It is here that body burdens are greatest and toxic effects most pronounced. In white-tailed sea eagles, as for Baltic seals, reduced reproductive success is well correlated with bioaccumulation of DDE and PCB (Helander *et al.*, 1982).

An analysis of the data on concentrations of DDT and PCBs in guillemot (*Uria algae*) eggs and in herring muscle show a decline over the period 1968–84 (Olsson & Reutergårdh, 1986). In Gulf of Bothnia herring, DDT concentrations have declined more rapidly than PCBs. Olsson & Reutergårdh (1986:108) concluded that with respect to toxic substances, the Baltic is "recovering more quickly than the most gloomy prophecies gave us reason to believe".

TEMPORAL SPREAD OF INCIPIENT PATHOLOGY

The history of the Gulf of Bothnia shows clearly a number of symptoms of incipient ecosystem pathology. The evidence for ecosystem breakdown at the 'local' level is incontrovertible. The appearance of abiotic zones, elevated primary productivity, profound changes in community structure and bioaccumulation of contaminants, all point to abnormalities in energy processing, nutrient cycling and a loss of information (*sensu* Margalef, 1985). At coastal levels, some ecosystem distress symptoms are also present, but these are less pronounced. At basin levels, again some clearer signs of pathology have been shown; for example the failure in reproduction of ringed and grey seals; the demise of the genetic diversity in salmonids; increases in nutrient concentrations in the open water and suggestive increases in secondary productivity in the macrozoobenthos.

All this suggests a rather complex pattern of pathology. Looking at particular areas (e.g. the Kyrönjoki Estuary) and the Gulf as a whole, four main patterns or stress gradients and discontinuities are to be found.

(1) The spread of pathology from local to coastal areas occurs via two pathways. One is found where a local disturbance, for example the effects from nutrient loading from municipalities, aggregates and becomes a regional problem as occurred in the Turku-Naantali area (Pitkänen *et al.*, 1987). A second is one in which a local disturbance affects major centres of organization. This was illustrated by the events in the Kyrönjoki River Estuary whereby local disturbance caused losses to critical habitat for bream and burbot and thus had the consequences of reducing their abundance over the entire migration area (20–30 km of coastal area).

(2) The spread of pathology from local to basin-wide areas. This is best illustrated by the destruction of spawning habitats of anadromous whitefish, salmon and sea trout, owing to a variety of causes (e.g. various hydro-engineering works). These local activities cumulated into a basin-wide decline in genetic diversity; over half the original stocks of anadromous salmonids and coregonids have been eliminated (Hildén *et al.*, 1982; Lindroth, Larsson & Bertmar, 1982).

(3) Spread of pathology from coastal to local areas. A possible example, although at this point it remains a somewhat speculative hypothesis, is that diffuse loadings of nutrients via agricultural run-off have led to coastal eutrophication which in turn has caused damage to local spawing grounds, particularly for Baltic herring (Aneer, 1985).

(4) The spread of pathology from basin to local areas provides a fourth pathway for the transmission of ecosystem pathology. One of the impacts of increased fishing pressure in the Baltic Sea has been a decline in the numbers of spawners in salmon rivers. The decline has been so severe that large areas in the rivers have become devoid of juvenile salmon (Jutila & Pruuki, 1987). As juvenile salmon appear to play a central role in structuring fish communities in the rapids (Ikonen *et al.*, 1987), their decline damages these local ecosystems.

DISCUSSION

Taking the appearance of the symptoms of 'ecosystem distress' as evidence for ecosystem pathology, there is much to suggest that pathology is widespread within the Gulf of Bothnia ecosystem and moreover that it is transmitted from one region to another via a variety of mechanisms.

This conclusion might be called into question given the uncertainty that surrounds the interpretation of many of the data sets owing to lack of statistical or analytical rigour. There is much to suggest that data on 'big waters', such as the Gulf of Bothnia, are far from ideal for purposes of establishing conditions and trends over broad regions; for this reason it is essential to consider changes in a number of indicators. Practically every data set, considered alone, might be challenged. Yet, taken together, the data gathered for this study tell a persuasive story. The evidence suggesting incipient pathology is mutually consistent thereby reducing the uncertainties in interpretation of isolated data sets.

I have viewed developments in the Gulf of Bothnia from the perspective of ecosystem 'medicine' (Rapport, 1983, 1984; Rapport, Regier & Thorpe, 1981); that is, I have focussed on symptoms of pathology without attempting to explicate the mechanisms that may have led to these conditions. In ecosystems, as in human medicine, it would appear that multifactorial etiology is the rule rather than the exception (Dubos, 1968) and conflicting theories as to causes of disease or ecosystem pathology are often reconciled when viewed in this light.

Take, for example, the sudden disappearance (or weakened condition) of fucus (*Fucus vesiculosus*) in many parts of the Archipelago Sea over an area of some 3500 km^2 (Ronnberg, Lehto & Haahtela, 1985). When fucus vegetation began disappearing in parts of the Archipelago Sea, the demise was at first taken as a sign of eutrophication, since it is well known (from studies in the Baltic proper and elsewhere) that eutrophication causes increased turbidity, thus reducing the light needed to support fucus at lower depths. However, with further studies a much more complex picture is emerging in which increased nutrient levels, increased biomass of filamentous algae, grazing of fucus by a large population of the isopod crustacean *Idotea balthica,* and destruction by sea ice all play a role (Haahtela, 1984). Further, natural oceanographic processes appear to have contributed also to increased nutrient concentrations in the surface layers (Kangas *et al.,* 1982). The disappearance and partial re-establishment of fucus (Ronnberg, 1985) are now viewed as a result of an unusual combination of stresses (i.e. multifactoral etiology), rather than solely eutrophication or solely natural oceanographic processes.

The difficulties in analysis are further compounded by the long lags between the onset of stress and ecosystem response, which may range from years to many decades or even centuries (Clark & Munn, 1986; Hilden & Rapport, 1989). For example, nutrient concentrations in surface waters in the Gulf of Bothnia have been increasing for decades. Further, there is now abundant evidence for eutrophication in local and coastal areas (Pitkänen *et al.,* 1987). However, in the open sea, evidence for increased primary productivity has not been clearly shown.

Tracking the state of health of the Baltic Sea requires the development of a monitoring strategy which takes into account the various pathways by which pathology develops and the time lags involved. Monitoring activities in the Gulf of Bothnia have concentrated on the open sea, and on local areas of concern. The coastal zone has been relatively neglected. However, it is the coastal zone that plays a critical role in the propagation of stress from local areas to the open sea, and therefore more focus on developments here may provide 'early warning' of incipient changes in the larger system.

Margalef (1975:239) hypothesized "All or most of the ways in which man interferes with the rest of nature produce coincident or parallel effects . . .". Applications of stress concepts to the ecosystem level (Barrett, Van Dyne & Odum, 1976) have provided a framework for probing this hypothesis (Rapport *et al.*, 1985; Harris *et al.*, 1988). Stress concepts also play a key role in guiding the development of national environmental statistics and state of environment reporting in Canada, Sweden and other countries (Rapport & Friend, 1979; Rapport & Regier, 1980; Bird, Friend & Rapport, 1983; Bird & Rapport, 1986).

ACKNOWLEDGEMENTS

I thank colleagues in Sweden and Finland who gave unstintingly of their time to explain the intricacies of the Baltic ecosystem. I am particularly indebted to A.-B. Andersin, E. Bonsdorff, H. Cederwall, S. Fonselius, S. Hansson, M. Hildén, P. Kangas, H. Lehtonen, E. Leppäkoski, E. Neuman, R. Parmanne, M. Perttilä, R. Rosenberg, P. Tulkki and A. Voipio for guidance during the early phases of this work, and to S. Swanberg for translations of Finnish and Swedish documents. P. Calow, E. Leppäkoski, S. Hansson, M. Hildén and T. Polfledt made valuable comments on earlier drafts. Special appreciation to Agneta Odén for kindly drafting Fig. 1 and to Annabella Elliot for valuable editorial suggestions. The support of Statistics Canada (Ottawa), Statistics Sweden (Stockholm) and the Institute of Marine Research (Helsinki) made possible an extended and most pleasant research sojourn to Sweden and Finland during 1985–87.

REFERENCES

ANDERSIN, A.-B., 1987. The question of eutrophication in the Baltic Sea: results from a long-term study of the macrozoobenthos in the Gulf of Bothnia. *Publications of the Water Research Institute, Finland, 68:* 102–106.

ANDERSIN, A. B., GRIMÅS, U., KANGAS, P., LEHTONEN, H. & LEPPANEN, J. M., 1987. Review of the biology of the Gulf of Bothnia. *Publications of the Water Research Institute, Finland, 68:* 96–101.

ANEER, G., 1985. Some speculations about the Baltic herring (*Clupea harengus membras*) in connection with the eutrophication of the Baltic Sea. *Canadian Journal of Fisheries and Aquatic Sciences, 42 (Suppl. 1)*: 83–90.

BARICA, J., 1982. Lake Erie oxygen depletion controversy. *Journal Great Lakes Research, 8:* 719–722.

BARRETT, G. W., VAN DYNE, G. M. & ODUM, E. P., 1976. Stress ecology. *BioScience, 26:* 192–194.

BENGTSSON, A. & MIETTINEN, V., 1987. Spinal deformities and concentrations of heavy metals and chlorophenols in the fourhorn sculpin (*Myoxocephalus quadricornis* L.) off the Finnish coast of the Bothnian Bay, *Publications of the Water Research Institute, Finland, 68:* 175–178.

BERGMAN, A. & OLSSON, M., 1986. Pathology of Baltic grey seal and ringed seal females with special reference to adrenocortical hyperplasia: Is environmental pollution the cause of a widely distributed disease syndrome? *Finnish Game Research, 44:* 47–62.

BIRD, P. M., FRIEND, A. M. & RAPPORT, D. J., 1983. *Methodology for Identifying Environmental Indicators for the State of Environment Report for Canada.* Background paper prepared for the Conference of European Statisticians, Informal Meeting on General Methodological Problems in Environment Statistics. Helsinki, Finland 2–5 May, 1983. HE 83-30968.

BIRD, P. M & RAPPORT, D. J., 1986. *State of the Environment Report for Canada.* Ottawa: Canadian Government Publishing Centre.

BRUNEAU, L., 1980. Pollution from industries in the drainage area of the Baltic. *Ambio, 9:* 145–152.

CEDERWALL, H., 1987. State of the Swedish coastal zone of the Gulf of Bothnia. *Publications of the Water Research Institute, Finland, 68:* 122–128.

CEDERWALL, H. & LEONARDSSON, K., 1984. Monitoring av mjuk bottenfauna i Bottniska viken. Rapport for 1983. Naturvardsverket, Rapport SNV-PM 68. [In Finnish.]

CLARK, W. C. & MUNN R. E. (Eds), 1986. *Sustainable Development of the Biosphere*. Cambridge: Cambridge University Press.

DUBOS, R., 1968. *Man, Medicine, and Environment*. New York: Praeger.

DYBERN, B. I. & FONSELIUS, S. H., 1981. Pollution. In A. Voipio (Ed.), *The Baltic Sea:* 351–380. Amsterdam: Elsevier Scientific Publishing Co.

ENCKELL-SARKOLA, E., PITKÄNEN, H., RUOHO-AIROLA, T., ERIKSSON, B., WIDELL, A. & AHL, T., 1984. (Loading of the Gulf of Bothnia in 1972–1982). *Kommitten for Bottniska viken, Arsrapport, 11:* 15–53. [In Swedish.]

GODRON, M. & FORMAN, R. T., 1983. Landscape modification and changing ecological characteristics. In H. A. Mooney & M. Godron (Eds), *Disturbance and Ecosystem: Components of Response:* 12–28. New York: Springer-Verlag.

HAAHTELA, I., 1984. A hypothesis of the decline of the bladder wrack (*Fucus vesiculosus L.*) in SW Finland in 1975–1981. *Limnologica (Berlin), 15:* 345–350.

HÄLLFORS, G., NIEMI, A., ACKEFORS, H., LASSIG, J. & LEPPÄKOSKI, E., 1981. Biological oceanography. In A. Voipio (Ed.), *The Baltic Sea:* 219–274. Amsterdam: Elsevier Scientific Publishing Co.

HANSSON, S., 1987. Effects of pulp and paper mill effluents on coastal fish communities in the Gulf of Bothnia, Baltic Sea. *Ambio, 16:* 344–348.

HARRIS, H. J., HARRIS, V. A., REGIER, H. A. & RAPPORT, D. J., 1988. Importance of the nearshore area for sustainable redevelopment in the Great Lakes with observations on the Baltic Sea. *Ambio, 17 (2):* 112–120.

HELANDER, B., OLSSON, M. & REUTERGÅRDH, L., 1982. Residue levels of organochlorine and mercury compounds in unhatched eggs and the relationships to breeding success in white-tailed sea eagles *Haliaeetus ablicilla* in Sweden. *Holarctic Ecology, 5:* 349–366.

HELLE, E., 1986. The decrease in the ringed seal population of the Gulf of Bothnia in 1975–84. *Finnish Game Research, 44:* 28–32.

HELLE, E., OLSSON, M. & JENSEN, S., 1976. PCB levels correlated with pathological changes in seal uteri. *Ambio, 5:* 261–263.

HELLE, E., HYVÄRINEN, H. & STENMAN, O., 1986. PCB and DDT levels in the Baltic and Saimaa seal populations. *Finnish Game Research, 44:* 63–68.

HESSLE, C., 1924. Bottenboniteringar i inre Ostersjon. *Meddelanden fran kungliga Lantbruksstyrelsen, 250:* 1–52. [In Finnish.]

HILDÉN, M., HUDD, R. & LEHTONEN, H., 1982. The effects of environmental changes on the fisheries and fish stocks in the Archipelago Sea and the Finnish part of the Gulf of Bothnia. *Aqua Fennica, 12:* 47–58.

HILDÉN, M., LEHTONEN, H. & BÖHLING, P., 1984. The decline of the Finnish vendace, *Corregonus albula* (L.), catch and the dynamics of the fishery in the Bothnian Bay. *Aqua Fennica, 14:* 33–47.

HILDÉN, M., KUIKKA, S., ROTO, M. & LEHTONEN, H., 1988. Differences in fish community structure along the Finnish coast in the Baltic Sea. International Commission for Exploration of the Seas, 1988/BAL; No. 15. (17p. Mimeo).

HILDÉN, M. & RAPPORT, D. J., 1989. The propagation of ecosystem pathology in multi-stressed environments. Submitted for publication.

IKONEN, E., AHLFORS, P., MIKKOLA, J. & SAURA, A., 1987. *Meritaimenen ja lohen elvyttaminen Vantaanjoen vesistossa*. Finnish Game and Fisheries Research Institute, Fisheries Division, Monistettuja julkaisuja 62. [In Finnish.]

JUTILA, E. & PRUUKI, V., 1987. The improvement of the salmon stocks in the Simojoki and Tornionjoki rivers by stocking parr in the rapids. International Commission for Exploration of the Seas, C.M. 1987/ M:25. (17p Mimeo).

KANGAS, P., AUTIO, H., HÄLLFORS, G., LUTHER, H., NIEMI, A. & SALEMAA, H., 1982. A general model of the decline of *Fucus vesiculosus* at Tvarminne, south coast of Finland in 1977–1981. *Acta Botanica Fennica, 118:* 1–27.

LARSSON, U., ELMGREN, R. & WULFF, F., 1985. Eutrophication and the Baltic Sea: Causes and consequences. *Ambio, 14:* 9–14.

LEPPÄKOSKI, E., 1975. Assessment of degree of pollution on the basis of macrozoobenthos in marine and brackish-water environments. *Acta Academiae Aboensis, Series B, 35:* 1–90.

LEPPÄKOSKI, E., 1979. The use of zoobenthos in evaluating effects of pollution in brackish-water environments. In *The Use of Ecological Variables in Environmental Monitoring*. Report PM 1151: 151–157. The National Swedish Environment Protection Board: Stockholm.

LEPPÄKOSKI, E., 1984. Introduced species in the Baltic Sea and its coastal ecosystems. *Ophelia, Suppl. 3:* 123–135.

LINDROTH, A., LARSSON, P.-O. & BERTMAR, G., 1982. Where does the Baltic salmon go? In K. Muller (Ed.), *Coastal Research in the Gulf of Bothnia:* 387–414. The Hague: Dr W. Junk, Publishers.

MAKINEN, A. & AULIO, K., 1987. *Cladophora glomerata* (Chlorophyta) as an indicator of coastal eutrophication. *Publications of the Water Research Institute, Finland, 68:* 160–163.

MARGALEF, R., 1975. Human impact on transportation and diversity in ecosystems. How far is extrapolation valid? In *Proceedings of the First International Congress of Ecology: Structure, Functioning and* extrapolation valid? In *Proceedings of the First International Congress of Ecology: Structure, Functioning and Management of Ecosystems:* 237–241. Wageningen, Netherlands: Centre for Agricultural Publishing and Documentation.

MARGALEF, R., 1985. From hydrodynamic processes to structure (information) and from information to process. In R. E. Ulanowicz & T. Platt (Eds), *Ecosystem Theory for Biological Oceanography. Canadian Bulletin of Fisheries and Aquatic Sciences, 213:* 200–220.

NEUHOLD, J. M. & RUGGERIO, L. F. (Eds), 1975. *Ecosystem Processes and Organic Contaminants: Research Needs and an Interdisciplinary Perspective.* Washington: National Science Foundation.

NEUMAN, E. & KARÅS, P., 1989. Effects of pulp mill effluent on a Baltic coastal fish community. *Water Science Technology* (in press).

ODUM, E., 1985. Trends expected in stressed ecosystems. *BioScience, 35:* 419–422.

OLSSON, M., 1987. PCBs in the Baltic Environment. In J. S. Waid (Ed.), *PCBs and the Environment, vol 3:* 181–208. Boca Raton, Florida: CRC Press, Inc.

OLSSON, M. & REUTERGÅRDH, L., 1986. DDT and PCB polllution trends in the Swedish aquatic environment. *Ambio, 15:* 103–109.

PALOSUO, E., 1966. ICC in the Baltic. *Oceanographic Marine Biology Annual Review, 4:* 79–90.

PARMANNE, R., LEHTONEN, H., HÄKKILÄ, K. & TULKKI, P., 1986. Effect studies off a TiO_2 industry in the Bothnian Sea (Baltic) 1984–1985. International Commission for Exploration of the sea. C.M. 1986/E:46.

PITKÄNEN, H., 1986. The regional distribution of some water quality variables in Finnish coastal waters. *Publications of the Water Research Institute, Finland, 62:* 105–120.

PITKÄNEN, H., TULKKI, P. & KANGAS, P., 1985. Areal variation and trends in the nutrient concentrations of Baltic waters around Finland. International Council for the Exploration of the Sea, Hydrography Committee: C.M. 1985/c:47. 16p.

PITKÄNEN, H., KANGAS, P., MIETTINEN, V. & EKHOLM, P., 1987. *The State of the Finnish Coastal Waters in 1979–1983.* Helsinki: Vesi-Ja Ymparistohallinnon Julkaisuja, 8.

RAPPORT, D. J., 1983. Ecosystem medicine. In J. B. Calhoun (Ed.), *Environment and Population: Problems of Adaptation:* 96–98. New York: Praeger.

RAPPORT, D. J., 1984. State of ecosystem medicine. In V. W. Cairns, P. V. Hodson & J. O. Nriagu (Eds), *Contaminant Effects on Fisheries:* 315–324. New York: John Wiley & Sons.

RAPPORT, D. J. & FRIEND, A. M., 1979. *Towards a Comprehensive Framework for Environmental Statistics: A Stress-Response Approach.* Ottawa: Statistics Canada.

RAPPORT, D. J. & REGIER, H. A., 1980. An ecological approach to environmental information. *Ambio, 9:* 22–27.

RAPPORT, D. J., REGIER, H. A. & THORPE, C., 1981. Diagnosis, prognosis and treatment of ecosystems under stress. In G. W. Barrett & R. Rosenberg (Eds), *Stress Effects on Natural Ecosytems:* 269–280. New York: John Wiley & Sons.

RAPPORT, D. J., REGIER, H. A. & HUTCHINSON, T. C., 1985. Ecosystem behaviour under stress. *American Naturalist, 125:* 617–640.

RAPPORT, D. J. & REGIER, H. A., 1989. Disturbance and stress effects on ecological systems. In B. C. Patten & S.-E. Jørgensen (Eds), *Progress in Systems Ecology: Contemporary Issues and Perspectives.* Vol. 1. Englewood Cliffs, N.J.: Prentice Hall (in press).

REGIER, H. A. & HARTMAN, W. L., 1973. Lake Erie's fish community: 150 years of cultural stresses. *Science, 180:* 1248–1255.

RIKKINEN, K., 1980. The Baltic's urban systems. *Ambio, 9:* 138–144.

RONNBERG, O., LEHTO, J. & HAAHTELA, I., 1985. Recent changes in the occurrence of *Fucus vesiculosus* in the Archipelago Sea, SW Finland. *Annales Botanici Fennici, 22:* 231–244.

SKOG S. & VARMO, R., 1980. Effects of pollution on the distribution of *Macoma baltica (L).* in the sea area of Helsinki. *Finnish Marine Research, 247:* 124–134.

THOMPSON, J. S., 1982. An epizootic of lymphomia in northern pike *Esox lucius* L., from the Aland Islands of Finland. *Journal of Fish Diseases, 5:* 1–11.

URHO, L. & HUDD, R. Outbreak of fin erosion after an oil spill and the possible use of fin erosion in environmental monitoring. Submitted for publication.

VOIPIO, A. (Ed.), 1981. *The Baltic Sea.* Amsterdam: Elsevier Scientific Publishing Co.

WESTMAN, K. & KALLIO, I., 1986. Endangered fish species and stocks in Finland and their preservation. Paper presented to EIFAC/FAO Symposium on Selection, Hybridization and Genetic Engineering in Aquaculture of Fish and Shellfish for Consumption and Stocking. Bordeaux (France), 27–30 May 1986.

WOODWELL, G. W., 1967. Radiation and the patterns of nature. *Science, 156:* 461–470.

Biological Journal of the Linnean Society (1989) *37:* 51–78. With 6 figures

The analysis of stress in natural populations

A. J. UNDERWOOD

Institute of Marine Ecology, Zoology Building, University of Sydney, NSW 2006, Australia

Populations usually persist despite environmental variations. Experimental analysis of responses to stress must include distinction between potential stresses (environmental perturbations that might not cause stress) and actual stress (phenomena that cause a response by the population). This is made difficult by large temporal fluctuations in abundances of many organisms. Monitoring can measure this variability but is insufficient to predict the potential impact of most stresses. Experimental analyses of stresses are also made difficult by differences among populations in their inertia (lack of response to perturbation), resilience (magnitude of stresses from which a population can recover) and stability (rate of recovery following a stress). These attributes of populations cause a range of responses to intermittent, temporary and acute (or 'pulse') stresses and to long-term, chronic ('press') disturbances. The timing, magnitude and order of stresses can cause different responses by populations. Synergisms between simultaneous or successive stresses can also have unpredictable effects on populations and cause complexity in interpretations of patterns of competition and predation. Experimental manipulations are needed to understand the likely effect of environmental disturbances on populations. The appropriate experiments are those designed to measure the effects of different types, magnitudes and frequencies of simulated stresses. These will be more revealing than the more common experimental analyses used to determine why and how observed changes in abundances of populations are caused by existing stresses.

KEY WORDS:—Stress – perturbation – inertia – resilience – stability – experimental design – prediction.

CONTENTS

0024–4066/89/050051 + 28 $03.00/0

INTRODUCTION: STRESSES AS A CLASS OF PERTURBATIONS

The detection and measurement of stresses in natural populations are beset with considerable difficulties because of the variance in time and space of many natural systems. A stress is an environmental change that causes some response by the population of interest (e.g. Pickering, 1981). Stresses are therefore part of a set of changes that occur in environments. For clarity, all such changes will be considered as 'perturbations', although the term has potential to cause confusion. It implies that the environmental change has been caused by some process external to a population (e.g. a forest fire, an oil spill). In this usage, perturbation is sometimes excessively restricted to occasions when deliberate changes are made by investigators to a population (Bender, Case & Gilpin, 1984) or to an environment. Yet other authors have identified as perturbations (or disturbances) any changes in numbers of a population, including those that are brought about by intrinsic, normal functions of the populations themselves. Thus, Sutherland (1981) identified the arrival of larvae of a marine species as a perturbation of the adult population, even though the arrival of the larvae is, in fact, the 'birth process' of such populations. Despite potential difficulties, 'perturbation' will serve here to describe any natural, accidental or deliberately induced changes in environments.

Apart from stresses (i.e. perturbations causing a response in a population), there are perturbations that cause no response. Sutherland (1981) defined these as Type I perturbations and defined stresses as Type II (causing a temporary change from which the population recovers) or Type III (causing a permanent or, at least, long-term change in a population). For reasons developed later in this paper, stresses will not be divided into Sutherland's Types II and III.

Finally, there are perturbations that exceed the possibility of measuring a response by a population—because the population and its habitat are destroyed. Thus, filling in a swamp to build houses, or clearing land for agriculture completely destroys the habitat of some populations so that they cannot respond—they are locally extinct. That local habitat is no longer available for future colonization. To distinguish them from less severe situations, these perturbations can be described as catastrophes (McGuinness, 1988).

Any examination of stress in natural populations must be part of a consideration of perturbations—including those that do not cause a response (and are thus not stresses). Catastrophes will not be further considered here (even though they may be important agents of natural selection).

Natural populations

This preamble sets the scene for examining what must be known or inferred about perturbations, and the responses of populations to them, in order to detect potential stresses before they become unacceptable and in order to be able to explain and predict the responses of populations to perturbations (including stresses). This discussion will centre on the sorts of information field biologists or environmental scientists might need about natural populations—those in the field that are not controlled (as in cultures of laboratory rats) nor managed deliberately (as in crops, farms and zoos). The distinction between natural and managed populations is arbitrary and ill-defined (any population in a national park is managed because the environment is managed). What matters is that, for natural populations, the levels of environmental heterogeneity in time and space are allowed to vary widely and are not managed so that they stay within exessively limited boundaries.

The primary variable of interest when considering stress and a population is the abundance of the organisms. Operationally, ecologists are much more usually concerned with densities (i.e. abundance per some sampling unit), at least partially because of difficulties with defining the boundaries of populations. For the purpose of this discussion, the existence of stress is determined by the existence of a response; that is, by some change in abundance (or density) of a population, following a perturbation. Other attributes of populations (growth-rates of individuals, fecundity, age of reproductive maturity, age-specific schedules of mortality, etc.) are important. They are, however, secondary, in the sense that knowledge of them is necessary to explain why a specific type or magnitude of perturbation is a stress, and why the response by the population is of a certain size or type. Investigation of these secondary variables also helps explain why and how populations recover (or do not recover) from stresses. Note that lack of response by the population to some perturbation does not imply that the perturbation has no effect on the individuals. For example, the perturbation of adding hydrocarbons to some population of meiofaunal invertebrates may cause no response in terms of abundances of the animals. For this population therefore, hydrocarbons are not a stress (they caused no response). The population's lack of response might, however, be due to large and complex physiological responses by each individual. Thus, individuals were stressed by the hydrocarbons, but their physiological responses were adequate to prevent mortality or any changes in reproduction so that the abundance of the population was unaltered.

The first, and often most formidable, task is to determine whether stresses exist and to be able to measure the direction, magnitude and consequences of the responses to stresses.

CHARACTERISTICS OF POPULATIONS THAT ARE RELEVANT TO STUDIES OF STRESS

Considerable confusion has surrounded the terminology of properties of populations that determine their responses to perturbations. Three aspects of populations are particularly relevant (although there is considerable variation in discussions of the number of relevant attributes there might be; see Westman, 1978; Westman & O'Leary, 1986). First is that property of a population that determines whether or not it will respond to (be stressed by) a particular size, type or frequency of perturbation. Then there are two properties that dictate whether a population can recover from a stress. Thus, there are two components of what Margalef (1969) called "adjustment stability"—the capacity of a system to recover. One is the rate of response after the stress has caused some change. The other is that attribute of a population defining how much stress can be tolerated and still result in eventual recovery.

Inertia

The first component is known variously as "persistence" (Margalef, 1969), "resistance" (Boesch, 1974; Connell & Sousa, 1983) and "inertia" (Murdoch, 1970; Orians, 1974). These terms should not be considered to be interchangeable. As pointed out by Connell & Sousa (1983) and Sutherland (1981), persistence is simply the observed property of a population that it continues to exist without changes in numbers. This will occur whether or not there is any perturbation and therefore subsumes and confounds situations where no events occur and where perturbations occur but have no effect. "Resistance" is also a confusing term to describe the situation when no response is made to a perturbation. The word implies, and is used widely by physiologists and ecologists (Rapport, Regier & Hutchinson, 1985; Selye, 1973) to mean that there is a response. The response is some change in the internal working of an individual organism (Selye, 1973), population or community (Rapport *et al.*, 1985) to cope with, alleviate or remove the stress. These are obviously responses to the perturbation which are measurable (which is how we know that the system has resistance). Consider the case of an assemblage in which an environmental perturbation is followed by the replacement of stress-sensitive species by less sensitive species that are ecologically similar. This has been called "congeneric homeostasis" (Hill & Wiegert, 1980) and is one proposed mechanism causing resistance by a community. Only if our understanding of assemblages of species must disintegrate into the metaphysical mumbo-jumbo of considering communities as superorganisms (Simberloff, 1980; Underwood, 1986a) is it reasonable to propose that disappearance of some species to be replaced by others is a *lack* of response to a perturbation! It is thus counter to common sense to use resistance to define a lack of response to perturbation. "Inertia" is probably the most utilitarian term for a population's lack of response (Murdoch, 1970; Orians, 1974; see Fig. 1).

The inertia of a population can, in theory, be measured according to objective, external scales of reference; that is, the maximal magnitude of a particular type of perturbation that causes no response (which is therefore equal to the smallest magnitude of that particular type of stress).

Stability

The second property of a population that enhances persistence despite perturbations is the rate of recovery after the stress appears. The problems of defining recovery (e.g. determining to what abundance the population will return after a stress is removed) will be considered later. For clarification of the relevant properties of species when stressed, consider a population that persists at

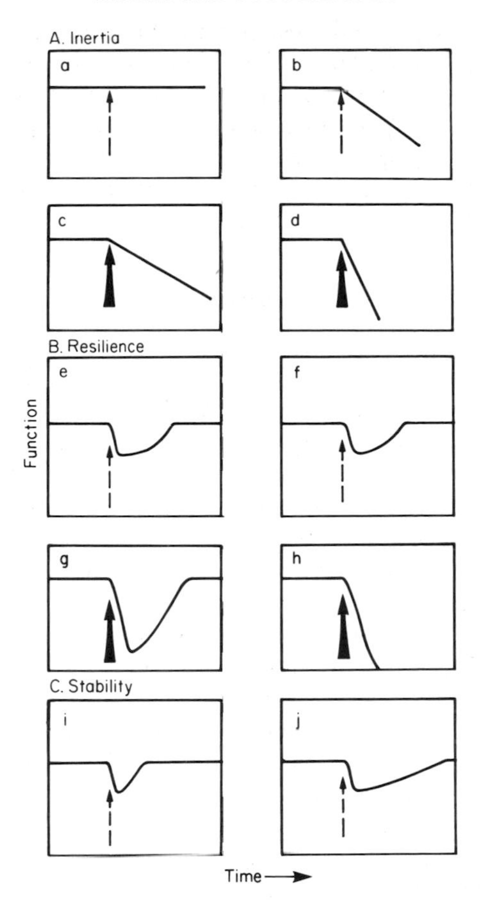

Figure 1. Diagrams illustrating characteristics of populations relevant to stresses for which an appropriate function is the abundance of a local population. Broken arrows represent a small perturbation of short duration; solid arrows represent a larger one. A, Inertia: the population on the left (a,c) is more inert than that on the right (b,d) and does not respond to a small perturbation (a), which stresses the other population (b); both populations are stressed by a larger perturbation (c,d). B, Resilience: the population on the left (e,g) is more resilient than that on the right; both can recover from a small perturbation (e,f), but only the population on the left (g) is sufficiently resilient to recover from the larger perturbation, which causes extinction of the less resilient population (h). C, Stability: the population on the left (i) recovers more quickly from a given perturbation than does the less stable population on the right (j).

constant abundance, or returns to the equilibrial abundance after being stressed (Fig. 1). If a population can recover (i.e. abundance returns to equilibrium) very quickly after a stress, the population can be considered 'stable' relative to a population that recovers more slowly. This use of stability to mean rate of recovery is inconsistent with other uses of the word, but should cause no problems provided its definition here is remembered. Other authors have used the term "elasticity" (Connell & Sousa, 1983; Orians, 1974) to mean rate or return to equilibrium. Elasticity can, however, also mean the *capacity* of an object to return to its original shape after a perturbation (i.e. non-elastic bodies never return to their equilibrial configuration). Thus, 'elastic' can also mean 'equilibrial' and confusion can still reign. Yet other authors have used the term "resiliency" (e.g. Boesch, 1974) to mean rate of recovery, but see below.

Whatever the term used, this property of populations is important. Note, however, that in contrast to inertia, stability must be measured in terms of a defined magnitude of stress. A population may recover at different rates (have different stabilities) with respect to stresses of different sizes. Knowledge of stability in response to some level of a particular type of stress may not provide information about stability in response to a different magnitude of the same type of perturbation. Often, it may only be possible to determine how stable a population is by comparison with another population (Fig. 1) or with reference to its stability when confronted by a different type or magnitude of stress.

Resilience

Finally, there is the attribute of a population that enables it to recover from different magnitudes of response to stress. This has been called "amplitude" (Orians, 1974) and "resilience" (Holling, 1973). Resilience is a better term to describe this property of a population. Amplitude should be reserved for the magnitude (size) of resilience of a particular population. The magnitude of resilience is measured by recording the amplitude of the largest response to stress from which a population can reattain equilibrium (see Fig. 1). Consider a population of equilibrial mean density 200 individuals per hectare. If the population can reattain this equilibrium (at whatever rate) after a stress large enough to remove 40% of the organisms, but cannot recover if 50% of the organisms are removed, its resilience is (i.e. has an amplitude of) between 80 and 100 individuals per hectare (see Fig. 1).

Persistence, perturbations and stress

This consideration of terminology leads to the following starting point for determining how to analyse the effects of stress on a population. First, the persistence of a population is its continuance through time at some particular (equilibrial) abundance, or its return to that equilibrial abundance after changes. Persistence may occur simply because there are no perturbations or because the population is sufficiently inert that it shows no response to the perturbations that occur. The inertia of a population therefore determines whether any particular magnitude and type of perturbation is a stress (i.e. causes a response). Persistence of a population at some equilibrium may also occur because the population can return to that equilibrium after responding to a stress. The rate of return after a response of a particular magnitude is the stability of the population. The maximal amplitude of response from which the population can restore its equilibrium estimates its resilience. This theoretical framework assumes that constancy of numbers (i.e. persistence and the equilibrium around which a population persists) can be identified over realistic spatial and temporal scales.

THE EXISTENCE OF STRESSES

Equilibrium and multiple equilibria

One of the major problems in the development of practical evaluations of stresses in natural populations is the determination of the existence of a particular stress—that is, detection that some particular perturbation has caused a response. If it were already established for some population of interest that its abundance was at some equilibrium and that the magnitude of the equilibrial abundance was already measured, the detection of stress would simply be the discovery that some perturbation had caused the abundance to be different from the equilibrial value. This is of course difficult, if not impossible, in nature, because populations do not exist in any sufficiently equilibrial numbers that changes in abundance in response to single perturbations are obvious, or likely. Many natural populations have considerable temporal and spatial variance in abundances so that the detection of a stress requires not that the abundance has changed at some place and time, but that the observed change in abundance is larger than can normally be expected to operate at a given place and time, given the processes that are already stressing the population.

There are great conceptual difficulties in definitions of equilibria in natural populations. The organisms and their food, predators and pathogens, fluctuate widely in time, and differ from place to place because of a host of ecological variables. Whilst these are usually responses to perturbations of some sort, they create a background against which the detection of equilibria becomes difficult, if not impossible. The conceptual and definitional problems have been well-reviewed by Connell & Sousa (1983) who proposed several criteria to determine that an equilibrium exists. Of particular relevance was the definition of timing of observations in natural populations relative to the longevity of the organisms. As already established by Frank (1968), equilibrial abundances of populations are not demonstrated simply by noting a lack of change from time to time. For very long-lived organisms, no change will be observed unless the population is stressed or unless the observations are over a period longer than the natural life-expectancy of the individuals. Consider a population of trees that have pulses of germination of sets of seeds and that usually live for two hundred years. If the abundances of the trees are monitored every year for, say, eighty years, and no changes in numbers are observed, this is not evidence for the abundance of the population being at some equilibrium. Demonstration of an equilibrial abundance requires that the trees are sampled at intervals longer than two hundred years. Then, if there is no temporal variation, the numbers are equilibrial, because the population restores itself to that equilibrium after the numbers change. Even where the time-course of sampling of a natural population has been attempted at suitably long intervals, there are few examples that indicate equilibrial abundances (Connell & Sousa, 1983).

There is also the problem that there is imprecision in sampling most species. It is unusual, to say the least, to be able to count all the members of a population. Consequently, abundances are estimated by some regime of sampling, leading to a probabilistic estimate of the mean abundance in some study area, coupled with an estimate of variance of that mean. The size of this variance (i.e. the degree of imprecision with which the true mean abundance is known) is a function of the intrinsic variability (dispersion) of the numbers of organisms per sample unit and the number of replicate samples examined (e.g. Andrew & Mapstone, 1987; Green, 1979). In consequence, even where a precise equilibrium might exist in the numbers of an organism, there will be differences from time to time in the estimated mean number, simply because of sampling error. Thus, some procedure is needed to determine that the amounts of difference from time to time are no larger than would be expected from the errors in estimation at any one time. Procedures for this are generally available (e.g. analysis of variance; Underwood, 1981), but care must be taken to ensure that sampling from time to time is sufficiently independent to satisfy the assumptions of statistical procedures (see the later discussion on this point). Keough & Butler (1983) have developed a procedure that can be used to determine whether the abundances of a species that have been estimated in a series of temporal samples have varied by more than would be expected from the intrinsic variation in sampling. The procedure requires several (at least six) independent censuses so that populations of long-lived organisms must be sampled over very long periods. Procedures of this sort also require some arbitrary definition of how much variation from time to time is accepted before the existence of an equilibrium is considered unlikely.

There is also a problem that numbers in a population can fluctuate stochastically around a limiting mean density without there being any operationally useful equilibrium. Such populations are "stochastically bounded" and unlikely to decrease to zero, nor to increase indefinitely (Chesson, 1978). There always exists an average abundance for any population. To demonstrate that the population is in equilibrium requires that the abundance is close to that average for a large proportion of independent times of sampling.

It has recently been proposed, at least in theory, that populations may take on several different equilibrial values—multiple equilibrial points. Thus, a population may oscillate or fluctuate about some equilibrial density then change and fluctuate about a much smaller or larger density for several generations (Holling, 1973; Lewontin, 1969; Sutherland, 1974). Connell & Sousa (1983) have examined the claimed examples of this phenomenon, but concluded that there is no evidence that any natural population has multiple equilibria. Apart from the problems of identifying a population that has changed from one equilibrium to another without the change being a response to some persistent stress (a 'press' phenomenon—see later), there are grave practical difficulties in sampling such populations and demonstrating different equilibria. Not only is there the problem of errors in estimating abundance at any one time, as discussed above, there is also the problem of distinguishing between two very similar patterns in abundance. First, if there really are multiple equilibria, the population should oscillate around one equilibrial value for several generations and then start to oscillate around another mean for several generations. The problem is to distinguish such a pattern from the alternative that the abundances are simply fluctuating in a random walk. The data plotted in Fig. 2 represent a population that varies in abundance by random decreases or increases at each time unit. It shows a pattern indistinguishable (by simply sampling) from that expected for a population varying from time to time from one equilibrium to another.

'Before' and 'after' contrasts of perturbed populations

Given the inherent theoretical and practical difficulties in being able to identify equilibria in natural populations, the obvious alternative is to determine whether the trajectory of mean abundance of a population is altered as a result of some perturbation. Two procedures exist for doing this, the first of which depends on knowing, in advance, that a particular perturbation is going to occur

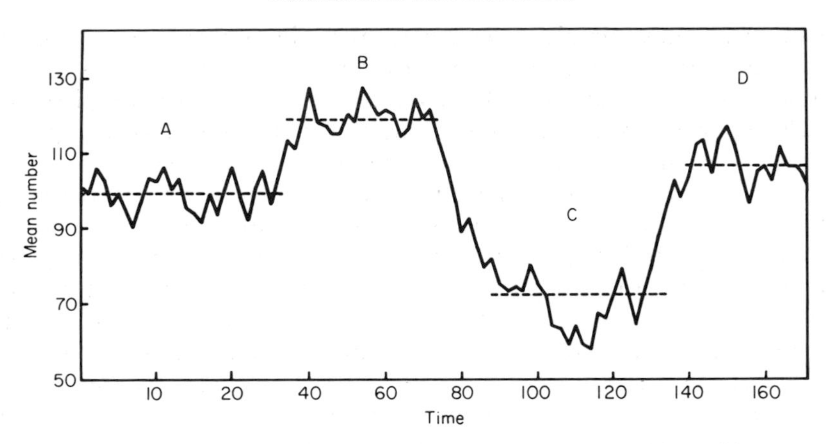

Figure 2. Apparent multiple equilibria in mean densities of a population, which oscillates around an equilibrium at A, then moves to a new equilibrium at B, and subsequently to equilibria at C and D. Data were, in fact, obtained from a computer simulation in which, at each time, a random number between -10 and $+10$ was added to the previous mean number. No process is operating to maintain the population at any of the apparent equilibria.

at some place and time. This procedure is a 'before' and 'after' comparison where the abundance of some population has been monitored for some period before the appearance of the perturbation. Detection of any change that might be due to the perturbation is then a comparison of the abundances after the perturbation with those before. If, as seems likely, the abundance of the population was varying before the perturbation began, the detection of stress (a response to the perturbation) is the identification that mean abundance after the perturbation has, in fact, become larger or smaller than the fluctuations recorded before the perturbation occurred. The logic of this comparison is that a stress would cause a change in the population leading to its abundance being different from those recorded in the absence of the perturbation—regardless of the constancy or equilibrial nature of the abundances before the perturbation started.

Such procedures are widespread, particularly in the literature on environmental impact, where a programme of monitoring is established before some environmental change (e.g. the construction of a power plant) and the monitoring continues after the perturbation (after the power plant begins operation). Any stress on a target species will be detected by changes in numbers after the plant starts (e.g. see the review by Green, 1979 and the recent discussion by Stewart-Oaten, Murdoch & Parker, 1986).

Despite its widespread use, this procedure is seriously flawed unless considerable spatial replication is involved. If monitoring is done at a single site before and after the perturbation, there is no method for determining whether any change during the period of perturbation is associated with the particular perturbation. For example, the numbers of a population of some species of plant may alter in a geographical region due to a general, regional change in weather. If this happened to coincide with the start of some perturbation of interest (e.g. the construction of a factory chimney) near to a site being monitored, the change in numbers would falsely be attributed to the effects of the perturbation. No single site can be expected to stay constant—even if the population happens to exist at some very close equilibrium during the 'before' stage of monitoring. To assume that the population would continue to be at that abundance in future, so that any changes can be associated with the single perturbation of interest, is an inductive conclusion beset with all the problems of inductive logic. Of course, most populations in nature are fluctuating in response to an array of perturbations, so that the most likely inductive guess about the future is that the mean and the level of fluctuation will both continue indefinitely at their present levels. This is not a satisfactory method for determining whether a perturbation has caused a response in the population.

Comparisons of perturbed and unperturbed populations

The second procedure for detecting stresses is to compare the abundances of a population in an area that has been or is being perturbed with that in a similar area that has not been affected. This is often done in an attempt to determine the effects of a potential stress once it has been observed (e.g. once an oil-spill occurs its effects are examined by investigations at the site of the spill in comparison to control areas).

The rationale behind such comparisons is that if the observed perturbation is a stress, the abundance of the target population in the perturbed area will subsequently differ from that in an area where no similar perturbation has occurred—regardless of any intrinsic spatial or temporal differences within the two areas.

Much of the literature on this topic is seriously flawed by spatial confounding—more recently called "pseudoreplication" by Hurlbert (1984). The problem is that a comparison of populations in any two sites, even though there may be considerable intensity of sampling and much spatial and temporal replication within the two sites, does not reveal why they differ. Thus, a difference between the two populations may be due to the perturbation or due to any other difference between the two sites that has nothing to do with that particular stress. Alternatively, it is possible that two sites differ naturally so that one has a smaller mean abundance of some species. If a stress acts on the area with greater abundance and reduces it to a similar mean as in the other area, a subsequent comparison will fail to detect the stress, without any 'before' information about the pre-existing difference.

In some studies, it has been suggested that this problem of confounding a particular stress with intrinsic spatial differences can be solved by comparing a perturbed and control site before and after the perturbation (Stewart-Oaten *et al.*, 1986). It is argued that a difference between the two areas in the magnitude of change before and after is evidence that the perturbation is a stress. Unfortunately, this does not solve the problem because it assumes that any pattern of difference between the two populations evident before the perturbation would have continued unchanged through time if the perturbation has no effect. This is the null hypothesis against which stress is to be detected. There is no logical reason to suppose that the original pattern will continue unchanged. Often, populations in two pieces of habitat have quite different temporal trajectories, regardless of any specific perturbation affecting one of them.

The only solution to the problem of spatial and temporal variance is to compare the populations in several, replicate perturbed and unperturbed habitats, or, at worst, to have several control sites even if there is only one perturbed site. There will still be all sorts of differences among a sample of randomly chosen control sites, and among randomly sampled perturbed sites. The mean abundance of a population and the changes in that population through time, when averaged over a sample of sites, will be the same from one whole sample (perturbed) to another (unperturbed) (Cochran, 1947; Fisher, 1932; Green, 1979; Hurlbert, 1984). Any difference in mean abundance of some species between the perturbed and control sites can then be attributed to the presence of the stress in one set of sites. All other, intrinsic, confounding variables have been averaged out.

Where only one perturbed site exists, its abundance, or change in abundance, through time should be compared with the average in a set of control sites. Only if the data from the single perturbed site fall outside the range of the control sites is it possible to argue that the perturbation is causing a change in (a stress to) the population.

Correlations and spatial comparisons

Another procedure that is sometimes used to detect stresses is to compare populations in a series of sites that have been perturbed to different extents. This is particularly widespread where toxins or other pollutants are believed to be affecting a population. The procedure is to estimate abundance of the population and the intensity, frequency, etc., of perturbation (usually the concentration of a putative pollutant) in a series of places and then to show that the abundances differ from place to place. The second step (Underwood & Peterson, 1988) is to demonstrate that the pattern of difference is associated with an increase or decrease in amount of perturbation. This is most commonly done by discovering a significant correlation between the abundance of organisms and the degree of stress.

This procedure is not without logical flaws because it does not unambiguously demonstrate that the pattern of difference among sites is caused by the perturbation. This could only be demonstrable if it were known that no other environmental variable had a similar trend from place to place as that measured for the perturbation. It is always possible that the pattern of correlation is caused by something else. For example, if some pollutant is released by an industry at the inland end of an estuary and is believed to be killing animals in the estuary, samples taken in sites from near the source of the pollutant and down to the mouth of the estuary will show a decreasing gradient of concentration of the chemical (a decreasing magnitude of perturbation). A negative correlation between the number of animals and the concentration of pollutant would be interpreted as evidence that the chemical is a stress. Of course, in such environments, there may also be gradients of increasing depth and salinity towards the mouth of an estuary. Either of these could be the cause of the differences in abundances of the populations.

Determination of causation in a pattern of correlation between abundance of a population and intensity of a perturbation is a difficult task and may only be possible by complex laboratory and field experimentation (see also Berge *et al.*, 1988; Underwood & Peterson, 1988).

The detection of a stress acting on a natural population is clearly fraught with difficulty because of the intrinsic spatial variability and lack of equilibria of real populations. The stresses most likely to be detected are those caused by planned perturbations (such as building a power-plant) so that a well-designed monitoring programme can compare populations around the site of the development with those in randomly-sampled control sites. Where perturbations occur dramatically and obviously (oil-spills, fires, floods, crop-spraying, etc.), their effects can be detected by appropriately replicated sampling and monitoring. The identification and detection of slow, chronic stresses is usually much more difficult because they have time-lags before their effects are apparent and because they require considerably more care in detecting them against background noise in the abundances of populations. If a perturbation is very widespread, it may affect all populations, making it impossible to determine causation in the field (there are no unperturbed control sites). Explanations of such stresses (and their subsequent management) are going to depend on laboratory studies on individuals and on suborganismal investigations.

MONITORING PROGRAMMES

A widespread activity in environmental management is to establish a monitoring programme, usually of long-term, to determine what changes (stresses) occur in a population or set of populations of interest. The programme, if well-designed, has the aims of estimating the natural rates of change in response to various forms of perturbation so that responses to particular (usually man-induced, accidental and unwanted) perturbations can quickly and unambiguously be detected. This distinguishes monitoring programmes from sampling done to describe the time-courses of abundances of populations. Monitoring is done with the twin aims of describing changes and of attempting to explain changes that are thought to be due to perturbations. There are, however, many problems with the use of monitoring, on its own, to detect anything of importance about a population.

Frequency of sampling

The first problem to confront the designer of a programme of monitoring is that the frequency with which samples must be taken is not easy to determine unless the sorts of time-scales of stresses and the responses to them by a population are already known. Where they are already known, there does not seem to be much point in establishing the sampling in the first place. Why the frequency of sampling is so crucial can be demonstrated by a simple example (Fig. 3). The abundance of the organisms varies in time because of natural declines, such as senescence of individuals, or because they are consumed by predators. At three times (identified by arrows), there are stresses (say due to storms) which remove considerable numbers of individuals of all ages. These are followed by episodes of recruitment of new individuals, and the numbers recruiting vary from one time to another. Thus, at any time, the population consists of several age-cohorts (indicated by different types of line in Fig. 3). Finally, consider that the individuals grow fairly rapidly so that their age is not

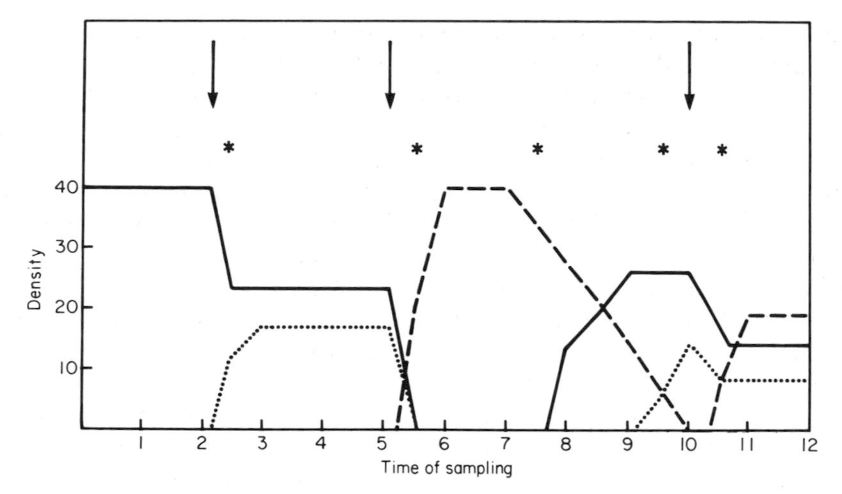

Figure 3. Diagram illustrating the problem of monitoring (i.e. sampling at regular times) a variable population. Different cohorts recruit into the population at various times, as indicated by different lines. Declines of density occur naturally and as a result of three perturbations, indicated by arrows. The summed density of all cohorts is 40, except for brief periods (marked by asterisks), which all occur between samples.

readily discerned by their sizes when they are being counted or sampled. The periodicity of sampling is indicated on an arbitrary time-scale.

In this particular (contrived) example, monitoring, however intensely done and however well-replicated, will not detect any of the changes in the population. There are only brief periods (all between times of sampling) when the density is below the original 40 per sample unit (these are indicated by asterisks in Fig. 3). In each case, the total density is very close to 40. The monitoring programme would fail to detect the natural rates of change (i.e. those not due to the particular type of stress being examined) and would completely miss the three actual stressful perturbations. In this case, the major reason for not detecting changes in numbers is that the recruitment of juveniles follows very rapidly upon the decline in abundance of previous adults. This is not an unlikely life history and applies to many weed species (Gadgil & Bossert, 1970; Harper, 1977) and others that have continuous reproduction, a short life-span and/or other properties of 'fugitive' species (Dayton, 1975; Hutchinson, 1951).

A monitoring programme is useful as an adjunct to experimental and more interactive studies of stresses. Very stable populations respond to, and recover from stresses very quickly. Knowledge of the stability of a population is required to plan an effective sampling programme. Where organisms respond to stresses at a rate slower than the rate of sampling, responses will be detected. It is, however, likely that such populations are also responding to many other perturbations at slow rates and the effects of a single type or episode of perturbation may be very difficult to identify. Some monitoring programmes have been successful in detecting stresses because they have been planned with the time-scale of the stress already well-known (e.g. Barnett, 1971; Hendricks *et al.*, 1974).

The design of a useful programme of monitoring must therefore take into account the rates of change of the species, the longevity, recruitment and other attributes that will affect stability and resilience. Populations that have little inertia to a range of perturbations will be exceptionally difficult to monitor for particular types of stress. They will change in abundance in response to a variety of perturbations and will thus have considerable temporal variance. Detection of a particular stress will therefore only be possible if it causes a response much greater than the background stresses operating on the population.

Thus, monitoring programmes are most likely to be useful for relatively inert, and not very stable populations. Even then, they should be coupled with manipulations simulating various stresses, and creating particular perturbations in order to determine the likely consequences to and responses by the organisms to environmental changes. Consideration of the inertia and stability of a population should be added to the list of relevant properties of species chosen for monitoring to detect environmental change in assemblages of species (see also Underwood & Peterson, 1988).

Spatial hierarchies of sampling

A second problem concerning monitoring programmes is the spatial scales at which monitoring is to be done. Clearly, if only a very small patch of a population's habitat is monitored, there will be scope for very large fluctuations in abundance (see Connell & Sousa, 1983). Rapid and continuous change can be expected, making it difficult to detect and identify the response to particular sorts of stress for which the population is being monitored.

Another difficulty with spatial scale of monitoring is that very localized stresses will only be detected if, by chance, they happen to occur in the sites being monitored. Local perturbations are unlikely to do this unless the monitoring programme is very intensively sampled at very many localities.

Conversely, if the population is sampled over very large spatial scales (e.g. over hundreds of miles of coastline), the laws of central tendency suggest that there will be little difference in mean abundance from time to time even when local or regional perturbations are common—unless perturbations are so massive and wide-scale that they alter the mean of the entire population over this geographic range (see Lewis, 1976, for some discussion of this point).

For most populations, it is not obvious what is an appropriate spatial scale for sampling. Unless populations are clearly delimited into natural patches (which then provide the spatial information necessary to determine the scale of sampling, and each patch becomes a unit of interest), it is important to use some form of hierarchical sampling strategy. Nested sampling designs are particularly useful because a series of arbitrarily chosen, but hierarchically inclusive, spatial scales can then be considered simultaneously without having to choose a particular scale or set of scales in advance (see Andrew & Mapstone, 1987; Green & Hobson, 1970; Underwood, 1981).

Time-lags and detection of stress

Another major problem with monitoring programmes is the potential existence of time-lags in the responses of organisms to stress. Some perturbations to abundances of populations take a long time to become evident, even though they may be short-term, acute perturbations causing relatively rapid responses at cellular, tissue or physiological levels of organization. Consider a stress such as that caused by some chemical continuously released into the environment that has only a very small effect on numbers in some population, but is accumulated and stored in the tissues of the individuals. When the concentration in the tissues is sufficiently large, the animals may cease to breed and abundances will decline drastically. By the time a monitoring programme detects the stress, it may be very difficult, if not impossible, to identify the cause. It is even possible that the perturbation has ceased by the time the response to it was detected. A long time-lag seems to have occurred with the pesticide DDT, which was in widespread use for twenty years before its deleterious effects on raptors were detected (Brooks, 1972; Hickey & Anderson, 1968).

Any perturbation that does not cause an immediate effect on the abundance of a population, because it is buffered by responses at lower grades of organization (physiological, behavioural, cellular, etc.) has the potential to cause very rapid changes once the stress becomes measurable as a change in abundance. For this reason, it is unlikely that monitoring of abundances of populations, on its own, is a very useful tool for detecting and predicting stresses in natural habitats.

Unfortunately, to link such monitoring to physiological, cellular or biochemical assay, depends on knowing that these aspects of the biology of a species are, in fact, predictive of future change in abundances (Underwood & Peterson, 1988). Thus, physiological measurements may indicate a response to some stress (such as pollutants in the environment). It might therefore be inferred that such physiological changes will decrease rates of reproduction and therefore reduce future abundances of the population. Whether this prediction is true depends on the degree of coupling between local rates of reproduction and future recruitment into the local population. For many marine species of invertebrates and fish, there is considerable uncoupling of reproductive output in some habitat and the subsequent rates and timing of recruitment, because of vagaries during the dispersive, planktonic 'mystery' stage (Spight, 1975). Predictions about future abundances of a population based on physiological measures are therefore unlikely to be of great accuracy. Only when the perturbation of interest is occurring over considerable proportions of the population's entire geographic range, or where local populations have relatively tight relationships between current reproduction and future size of population will it be possible to detect slowly-lagged perturbations well in advance.

Utility of monitoring

Monitoring programmes are therefore only of value to estimate natural rates of change of populations. If they are designed properly, variability in time can be estimated at several spatial scales. This provides information against which to detect the existence of a stress, but offers no useful information to predict what may happen once a stress occurs. The variations in numbers caused by stresses may be very different from those occurring under unstressed conditions (Hilborn & Walters, 1981). Monitoring programmes are unlikely to detect many sorts of stresses that are short-term or local and there is no way to anticipate exactly what, where and when to monitor because the nature and effects of stresses are unknowable in advance (Bisset & Tomlinson, 1983; Hilborn & Walters, 1981).

Only where a perturbation is deliberate and planned in advance is it possible to construct an appropriate monitoring programme to determine how much stress it causes. This type of monitoring, after an experimental or man-made perturbation, was discussed earlier.

There is also the problem of determining, in advance, which species to monitor. It may prove too expensive, time-consuming or taxonomically difficult to concentrate on rare species, which are the majority in most habitats. Thus, it is sometimes advocated that attention should be focussed on the abundant and widespread ones (e.g. Lewis, 1976). Yet these may be the least likely to show rapid responses to perturbations, either because they are well-adapted to a wide range of environmental conditions or because they have great and rapid powers of dispersal and recolonization after being perturbed. These are the characteristics that tend to make a species widespread and abundant. Because monitoring programmes are expensive and labour-intensive, the choice of species to study is crucial (Underwood & Peterson, 1988).

EXPERIMENTAL PROCEDURES FOR UNDERSTANDING STRESSES

To be able to understand, explain and predict the effects of stresses on populations, it is necessary to be able to estimate the inertia, resilience and stability of those populations with respect to appropriate types of perturbation. This is only possible by some experimental programme. Field experiments have been remarkably successful in elucidating the mechanistic processes acting on populations in complex and variable environments (e.g. Connell, 1972; Paine, 1977; Underwood, 1985, 1986b).

It is relatively easy to simulate processes that might remove different numbers of animals from (or add to) a population (see examples in Underwood, 1986b). The different manipulations represent various intensities or magnitudes of stresses. Alternatively, different frequencies of stresses can be examined by repeated experimental manipulations (Baker, 1971). What is needed is a change of emphasis in experimental studies of stress. The current major focus is to design experiments so that the observed effects of existing or previous stresses can be explained. For example, much of the experimental work on the effects of pesticides on birds was done retrospectively to explain why and how pesticide residues caused declines in abundances of raptors (Brooks, 1972; Hickey & Anderson, 1968). A quite different type of experiment is needed—designed to estimate the possible effects of different types of perturbations before they happen. To understand the most important aspects of the biology of populations in response to stresses, field experiments should be designed to estimate the inertia, resilience and stability of populations of different starting abundances in response to various experimental perturbations.

Use of such experiments would provide unparalleled information about the resilience and stability of populations under different types of stresses. More subtle manipulations could identify processes of stress and responses to them when specific parts of a population are removed (i.e. experimental harvesting or fishing) or added. The information obtained would allow considerably accurate predictions about the consequences of stresses not yet seen in nature. If the response of a population to the systematic removal of, say, 30% of the organisms were known, the consequences of any perturbation likely to cause a reduction of 30% or so in abundance would be known in advance.

Experimental perturbations

The need for appropriate controls and replication has been discussed above, but still seems to cause difficulties in the planning of field experiments (see particularly Hurlbert, 1984; Underwood, 1986b). These fundamental components of experimental procedures will not be considered further here. Instead, some of the particular problems of designing experiments on perturbations of populations will be considered.

Independence of sampling

Before results from field experiments can be analysed, it is important to make sure that data have been independently sampled in space and time. Independence of replicates and among experimental treatments is a fundamental assumption of most statistical procedures (Cochran, 1947; Green, 1979; Snedecor & Cochran, 1980; Underwood, 1981). Yet is has been violated in many studies of populations. From the viewpoint of analyses of stress, non-independent sampling through time is probably the most serious problem. Spatial non-independence will not be considered further.

The most obvious cause of temporal non-independence is sampling fixed areas, quadrats, positions, etc., at successive times. This has two effects on the data. Consider the simplest situation, such as a population of plants being monitored every two months. Ten quadrats have been randomly chosen in the study area at the beginning. Consider that, by chance, the quadrats sampled happen to be unrepresentative of the original population (because of sampling error) and are biased by being in locations that happen to have more (or less) robust plants than is the case on average in the study-area. Because these are repeatedly sampled on subsequent occasions, the original bias will continue throughout the entire set of data. Thus, the rate of change in the population will be considerably underestimated (more robust plants) or overestimated (less robust) in the samples. Where independent samples are taken, new randomly-chosen quadrats are examined at each time and it is extremely unlikely that these will repeatedly be biased. Independent sampling is more likely to be accurate (unbiased) and truly representative of the population as a whole thoroughout the period of study.

Second, and equally importantly, a series of non-independently sampled quadrats causes great biases in statistical analyses of the data. In analyses of variance (or their non-parametric equivalents) to test for temporal variations, non-independently sampled quadrats cause consistent underestimation of differences among times (unpublished data and McGuinness, personal communication).

Consider an experiment to determine the effects of a stress. A number of experimental plots, areas or quadrats is chosen and half of them (at random) are perturbed. If the organisms in the two sets of quadrats are then counted on successive occasions to detect any change in abundance due to the perturbation, the data are not independent through time. A stress would be revealed by the presence of an interaction between treatments and time (Green, 1979), because the perturbed population should respond differently (have a different temporal trajectory) from the control. Where repeated counts are made on the same plots (quadrats), any statistical analysis (e.g. analysis of variance, or comparison of regression lines of numbers on time, etc.) will overestimate the significance of such an interaction and therefore overestimate the existence and magnitude of response to stress.

The only solution to this problem is to have sufficiently large experimental plots that they can be randomly sub-sampled at each time period with independently drawn samples.

Press and pulse experiments

Two different sorts of stresses are possible. A 'pulse' stress is a relatively sudden, short-term and possibly acute stress which is then alleviated. Recovery by the population, provided it is sufficiently resilient, can be expected. 'Press' stresses, however, are sustained, long-term chronic perturbations and are likely to cause quite different patterns and time-courses of response (Bender *et al.*, 1984). For example, sustained stress due to fishing usually removes the larger members of inshore invertebrate fisheries. The abundance of a population may, at first, decline, but the population may subsequently show a progressive shift to similar abundances of smaller-sized individuals. Press stresses due to introduced, exotic competitive species might cause an indigenous population to become locally extinct in some patches, but remain at the previous density in parts of the habitat where it can resist invasion. The overall abundance will decline, but spatial variance among parts of the habitat would be increased in response to the stress.

Theories of press and pulse experiments, in so far as they relate to reductions in density of a population, have recently been reviewed (in a different context) by Bender *et al.* (1984). Two major differences in the effects of these two types of stress are important here. First, there is a fundamental difference in what can be measured. Pulse experiments are more prone to experimental error because the rate of responses by the perturbed population is more difficult to measure than is the magnitude of a sustained response due to a press experiment. Second, press experiments are of indefinite time-course and, unless an unambiguously detectable new equilibrial abundance is reached by the experimental population, it is not clear how to determine when the population has responded and the results of the experiment are known. This latter distinction is more apparent than real (see the discussion of convergence experiments below).

Time-courses and convergence experiments

One major problem with the design of any experimental manipulation or perturbation of a population is that of determining when the experiment has been completed. A typical pulse experiment consists of a sudden reduction (or increase) in abundance. Resilience is measured by noting what magnitude of response occurs and from which the population can 'recover'. Stability at that magnitude of stress is estimated by the time taken for the population to 'recover'. In each case, 'recovery' must be carefully defined. Previous discussion indicated

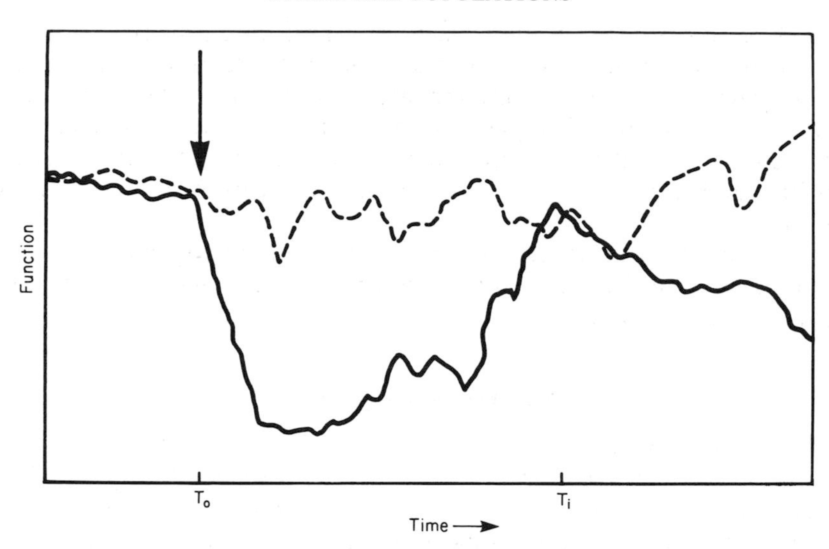

Figure 4. A convergence experiment to examine recovery of a population after a stress. The experimental population (solid line) was stressed at time T_o (arrow); the control (dashed line) was not perturbed. The function monitored (mean density) apparently converged at time T_i, but the two populations subsequently diverged and recovery was not complete (see text for further details).

that it was unlikely that any population would return to its previous abundance after manipulation. Recovery must therefore be defined as the mean abundance of replicated, manipulated populations becoming no longer different from that of replicated controls, regardless of the abundance to which they converge (see Bender *et al.*, 1984; Pielou, 1974; Underwood, 1986a).

The problem is that oscillating populations may converge and then subsequently diverge again, because recovery is not yet complete. This is illustrated in Fig. 4 where the mean of an experimental population converges with the mean of the controls, but recovery is not complete. If the experiments stopped at the first point of convergence (T_i in Fig. 4), the resilience and stability of the population would be overestimated. If possible, therefore, the projected time-course for convergence must be specified in advance of the experiment. Convergence by a specified time would be used in the definition of resilience and stability.

Alternatively, the experiment could be terminated if similarity of control and experimental populations has persisted for some previously specified period. Either strategy is going to require considerable thought and knowledge of the appropriate life history of the organisms before the results could be interpreted meaningfully.

Non-linearities in responses to stresses

The last problem considered here in the planning of experimental stresses is that responses may be non-linear (Fig. 5). The manipulation of a population by subjecting it to a single magnitude of a chosen type of perturbation may cause no response. A hypothetical case is illustrated in Fig. 5, where a population exists under a small, natural amount of stress (the Control) and a large stress is

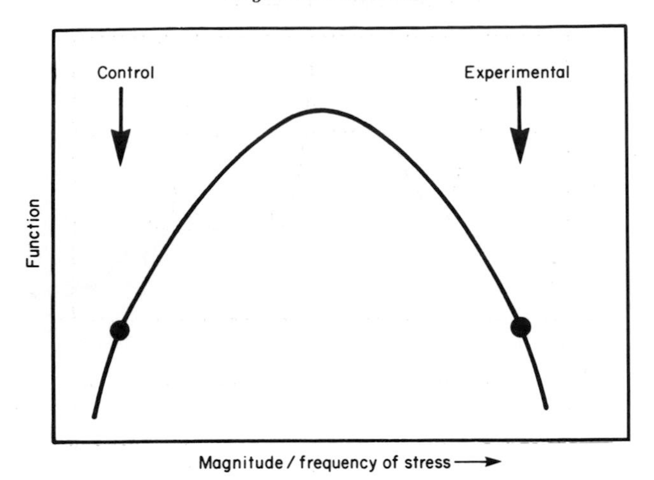

Figure 5. Non-linearities in response to a stress. The function monitored (e.g. density of a population) responds parabolically to increasing stress. Comparison of a Control population with a population perturbed experimentally (by the amount indicated for Experimental) would reveal no response to the perturbation.

imposed experimentally (the Experimental). Throughout the experiment, the function measured (in this case, abundance) is not different between the two treatments, suggesting that the population is inert to this perturbation. Yet, at smaller intensities or frequencies of perturbation, the population would have shown an increase in abundance (and therefore not have been deemed inert). Real examples of this occur in some plants that are affected by fire so that reproduction is enhanced by a minor stress due to fire, but excessive fire would destroy the seed-bank and no increased reproduction can occur. Such non-linearities are also discussed in Green (1979) and Underwood (1985). Experimental evaluations of the effects of stress should involve a range of rates or magnitudes of putative stresses so that non-linear responses by populations will be detectable (Underwood, Denley & Moran, 1983).

SOME CONSEQUENCES OF CHARACTERISTICS OF STRESSES AND OF POPULATIONS

Frequency of stresses

If the inertia, stability and resilience of a population were known with respect to various types of stresses, the consequences of stress to the population should be, to some extent, predictable. One obvious problem for persistence of relatively unstable populations is that further stresses may arrive before recovery from a previous one has been completed (Fig. 6). Thus, a population may be reduced to very small abundances by a series of stresses and, unless it is also very resilient, there is every chance of extinction because densities are so small relative to the potential carrying capacity. Thus, populations slow to respond to stresses should, in general, also be highly resilient. Incidentally, repeated stresses at an interval faster than the stability of the population creates a pattern of stress part-way between Bender *et al.*'s (1984) dichotomy of press and pulse phenomena.

It is also possible that individuals within a perturbed population are stressed at the physiological or other level without showing any response in terms of

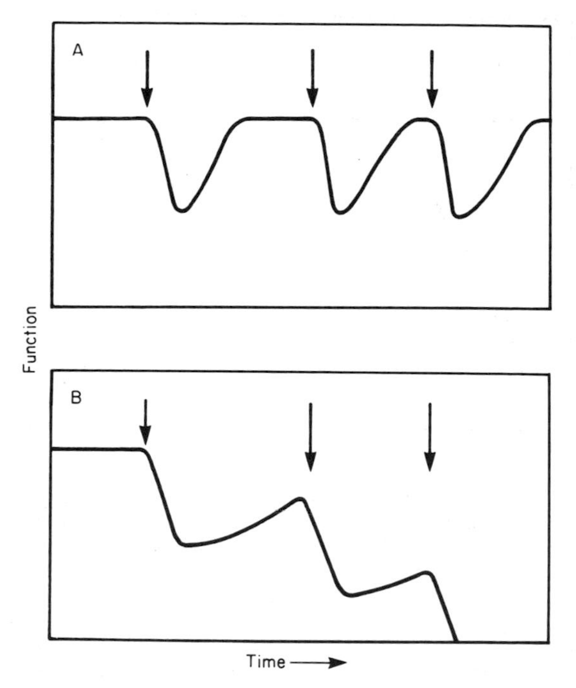

Figure 6. Consequences of repeated stresses to A, a stable and B, a less stable population. The abundance (the function plotted) of B does not have time to recover between stresses (arrows) and the population cannot persist.

abundance. Repeated stresses are then likely to exceed the suborganismal mechanisms of recovery and to start to kill individuals (e.g. Baker, 1971; Dimbleby, 1978; Loucks, 1970). As a result, the inertia of a population to any particular size or type of perturbation may be a very variable function of previous history of perturbations.

Timing of stress relative to the life cycle

If stresses occur at particularly crucial times of the life cycle, they may have far greater consequences than at other times. For example, Kennelly (1987) has demonstrated that the experimental removal of adult kelps from patches in a sublittoral forest has little effect if done at the time of year when the plants are breeding. If, however, the same stress, leading to the same response (i.e. the same reduction of density of adults) is done at other times of the year, there is no recovery until the next period of reproduction. In the interim, however, other species may occupy the resources (in this case space) and pre-empt them, making recovery impossible without further perturbation of the system.

Clearly, organisms may be adapted to stresses, and have evolved stress-related responses to enhance stability and resilience. The pattern of heathland plants in Australia responding to fire by germination of the seed-bank (and the converse that the seeds will often not germinate unless they have been burned) is clearly such an adaptation (Ashton, 1956; Gilbert, 1959). Yet other species have apparently responded to variable and frequent stresses by adopting a fugitive pattern of life history, where they grow quickly, reproduce quickly and then die, but their propagules are continuously available for recolonization of disturbed habitats (e.g. Gadgil & Bossert, 1974; Hutchinson, 1951).

If a stress occurs at the wrong stage of the life cycle for organisms well-adapted to stresses at other times, there may be very deleterious effects. If a forest is burnt when trees are young, there are much larger reductions in abundance than when the plants are older and are individually more robust (Heinselman, 1973).

It is also possible that a perturbation may arrive and decimate a population because most of the organisms were old, senescent or moribund and thus exceptionally susceptible to the stress (Heinselman, 1973). Consequently, the timing of the stress with respect to processes of reproduction, and with respect to the existing age- or size-structure of a population may be crucial in determining the outcome for the population.

Dispersive species

Widely dispersing species (many marine invertebrates and fish, many plants and insects) are more likely to be able to recover from localized stresses because the propagules of unaffected populations in other localities will recolonize an area. This has been argued as the main function of dispersal in many organisms—those that have persisted are dispersive (e.g. den Boer, 1979, 1981; Scheltema, 1971; Underwood, 1974). It is reasonable therefore to propose that widely dispersive and highly fecund populations are likely to have smaller inertia to a range of environmental perturbations than is the case for less dispersive, less fecund organisms.

It is nevertheless true that perturbations may not simply alter the abundances of a population, but may also alter the habitat in some way, making it unsuitable for continued occupation by a species. Then, only persistence on some regional, rather than local, scale is possible (Roughgarden, Iwasa & Baxter, 1985). Again, however, dispersive species are the only ones that can recolonize new habitat when perturbations make it available.

Rare species

Rare species pose some particular problems for interpretations of stress. It is not clear whether they are kept rare by sustained press stresses or whether the populations are kept sparse by small stability and resilience and numerous, repetitive and varied pulse stresses. Some rare species occasionally undergo outbreaks (e.g. insects, White, 1974, 1976; house-mice in Australia, Newsome, 1969; Newsome & Corbett, 1975), indicating that when long-term, press stresses are alleviated in nature they can respond very rapidly and build up very large populations. Rare species are not always well-adapted (i.e. by long-term patterns of natural selection) to such a life history. House-mice in Australia have only been there for relatively few generations (less than two hundred years) and it is not likely that they have altered much in such a short period (Newsome, 1969). Rare species are, however, subject to certain 'founder effects' in that local populations are often so small that genetic change may be very rapid but extinction in response to stress is also very likely (Simberloff, 1986).

Rare species may have non-linear responses to stress. When numbers of individuals are already small, a further small reduction may suddenly make finding a mate impossible and therefore the population will show very large consequent reductions. This is unlikely to be a problem for most species that are originally more abundant. It would be interesting to know how many rare species are rare as a result of identifiable press stresses (so that their conservation and continued management could be based on sound patterns of research aimed at identifying and eliminating these sources of stress).

Rare species are often ignored in monitoring programmes and therefore their responses to stressful and other perturbations may not be as well known as for more common and widespread species.

INTERACTIVE PROCESSES MAKING ANALYSES OF STRESS DIFFICULT

Synergisms and antagonisms

There are often complex sets of physical and biological processes acting together to determine the abundance of any local population. As a consequence, the effects of a particular type or magnitude of stress may be unpredictable, even if much experimentation has already revealed the nature of the population's responses to perturbations of a similar size and type. For example, any population that has just survived one type of stress may be much more vulnerable to the effects of a second, different type. Again, the inertia of a population to a given stress may be a function of the population's recent history. Two stresses acting simultaneously may have enhanced effects compared with those predicted from each alone (Hutchinson, 1973). Synergisms of this sort are an important component of any attempt to manage or conserve natural populations (Harris *et al.*, 1982).

Alternatively, the surviving individuals in a population after one stress may be those sufficiently robust to survive a second perturbation so that the population shows no further change in abundance when the second potential stress arrives. An example is that phytoplankton in a lake in Ontario were less deleteriously affected by the stress of heavy metals in the water when sewage (a second stress) was also present (Whitby *et al.*, 1976). Because this sort of antagonism (i.e. the effects of a second perturbation are eliminated by the existence of the first one) may occur, a proper predictive theory about the effects of stress must await considerably larger programmes of field experimentation, so that stresses acting in conjunction can be examined under natural conditions.

Density-dependence and independence

Another aspect of previous history that affects a population's response to stress is the sort of synergism between intrinsic processes acting continuously on the population and the arrival of a perturbation. For example, storms and other types of inclement weather are usually considered to act in a density-independent manner in populations (Andrewartha & Birch, 1984). This is not always the case, however, as demonstrated in experiments by Peterson & Black (1988) who found that bivalves at large density, which were already prone to density-dependent processes of competition, were much more subject to mortality when the sites were covered with silt (as during a storm) than was the case for bivalves at smaller densities. In this case, the effects of a storm would be density-dependent because of previous synergistic interactions within the population. Such synergisms make the effects of any perturbation very unpredictable without knowledge of other recent stresses and of previous processes that have led to the population having any given current inertia.

It has also been argued that insects are very much more damaging as herbivores of trees when the trees are already stressed by shortage of water (White, 1978) or air pollution (Dahlston & Rowney, 1980). As a consequence, the ultimate effects of a drought on a population of trees are going to depend on the numbers and diversity of insects that happen to survive the drought (or pollution) in any area. Yet again, the effects of any given perturbation are only going to be predictable, or interpretable, if considerable prior natural history and mechanistic understanding of processes affecting the ecology of a species are well understood.

Position in food-webs

Another component of the ecology of a species that makes the effects of stresses difficult to interpret concerns its position in a food-web. For example, if a storm (the stress) removes some individuals of a population (i.e. it is a relatively small stress), but decimates the food species of that population, the indirect effects of the stress are going to be very severe. In contrast, the storm may cause serious declines in the natural enemies of a population, making it more likely to increase in abundance, thus changing the eventual sign of effect of the storm from negative to positive.

Even this is not the potential end of indirect effects. There are examples of predatory species that are kept in check by press stresses due to the weather (Connell, 1975). A particular example is predatory whelks which eat mussels and barnacles on rocky shores of the east coast of the United States. The abundances or effectiveness of the predators are reduced by stress due to waves (which in some habitats is fairly continuous and therefore a press phenomenon; Menge, 1976, 1978). As a result, barnacles are affected by the stress and show an increase in abundance compared with habitats where wave-stress does not occur. So far, therefore, the effect of the stress on the barnacles is to cause an increase in density, because the predators are removed. Mussels now reach sufficiently large densities that they can overgrow and smother the barnacles, causing their densities to be maintained in small numbers (Menge, 1976, 1978). Unravelling such interactive processes is complex and unlikely to be effective without field experiments to test hypotheses about specific mechanistic processes that maintain abundances in different habitats.

CONCLUSIONS: MANAGEMENT AND REHABILITATION

The overwhelming conclusion that must be reached about the effects and consequences of stresses on natural populations is that they will only be predictable and explainable when the natural history of the organisms and their interactions with other organisms are well understood. The latter, in particular, usually requires fairly sophisticated field experimentation (look at the complex examples for only a few species on rocky shores in Underwood *et al.*, 1983 and subsequent variations on this theme in Dungan, 1986; Jernakoff & Fairweather, 1985; Lively & Raimondi, 1987; Sutherland & Ortega, 1986). Specific experimental programmes to investigate perturbations at different magnitudes and rates are needed for many habitats before the effects and consequences of stresses can be predicted with any reliability.

As ecology moves further from description and explanation to experiment and prediction, the processes of conservation and management will be much better based on scientific, ecological understanding. One current challenge is to unravel the patterns of inertia, stability and resilience in response to different types of stresses acting singly and in conjunction in different habitats.

One ultimate aim of this fundamental ecological research will be to identify better procedures for predicting and alleviating the effects of press stresses such as caused by pollution. In contrast, the consequences of more unpredictable pulse stresses will need to be much better understood so that programmes of repair and rehabilitation of natural assemblages and habitats can be attempted after these unpredictable stresses have occurred. Both tasks will progress much more rapidly if better designed field experiments were done to test more coherent articulated hypotheses about the processes and outcomes of stress.

ACKNOWLEDGEMENTS

The preparation of this paper was supported by funds from the Australian Research Grants Committee, the Marine Science and Technologies Grants Scheme of Australia, the Research Grant and the Institute of Marine Ecology of the University of Sydney and from Caltex (Australia) Pty. Ltd. I am grateful for the advice of and discussion with Drs P. G. Fairweather and K. A. McGuinness, and for the help and encouragement of Professor D. T. Anderson, Dr P. A. Underwood and Ms K. Astles. Comments from Professor P. Calow improved the manuscript.

REFERENCES

ANDREW, N. L. & MAPSTONE, B. D., 1987. Sampling and the description of spatial pattern in marine ecology. *Oceanography and Marine Biology: Annual Review, 25:* 39–90.

ANDREWARTHA, H. G. & BIRCH, L. C., 1984. *The Ecological Web.* Chicago: University of Chicago Press.

ASHTON, P. H., 1956. Speciation among tropical forest trees: some deductions in the light of recent evidence. *Biological Journal of the Linnean Society, 1:* 155–196.

BAKER, J. M., 1971. Successive spillages. In E. B. Cowell (Ed.), *Ecological Effects of Oil Pollution:* 21–32. London: Elsevier.

BARNETT, P. R. O., 1971. Some changes in intertidal sand communities due to thermal pollution. *Proceedings of the Royal Society of London, Series B, 177:* 353–364.

BENDER, E. A., CASE, T. J. & GILPIN, M. E., 1984. Perturbation experiments in community ecology: theory and practice. *Ecology, 65:* 1–13.

BERGE, J. A., SCHAANNING, M., BAKKE, T., SANDOY, K., SKEIE, G. M. & AMBROSE, W. G., 1988. A soft-bottom, sublittoral mesocosm by the Oslofjord: description, performance and examples of application. *Ophelia, 26:* 31–54.

BISSET, R. & TOMLINSON, P., 1983. Environmental impact assessment, monitoring and post-development audits. In PADC EIA and Planning Unit (Eds), *Environmental Impact Assessment:* 405–425. Boston: Nijhoff.

BOESCH, D. F., 1974. Diversity, stability and response to human disturbance in estuarine ecosystems. *Proceedings of First International Congress of Ecology (PUDOC), 1:* 109–114.

BROOKS, G. T., 1972. Pesticides in Britain. In F. Matsumura, G. M. Boush & T. Misato (Eds), *Environmental Toxicology of Pesticides:* 61–114. New York: Academic Press.

CHESSON, P. L., 1978. Predator-prey theory and variability. *Annual Review of Ecology and Systematics, 9:* 323–347.

COCHRAN, W. G., 1947. Some consequences when the assumptions for the analysis of variance are not satisfied. *Biometrics, 3:* 22–38.

CONNELL, J. H., 1972. Community interactions on marine rocky intertidal shores. *Annual Review of Ecology and Systematics, 3:* 169–192.

CONNELL, J. H., 1975. Some mechanisms producing structure in natural communities: a model and evidence from field experiments. In M. L. Cody & J. M. Diamond (Eds), *Ecology and Evolution of Communities:* 460–490. Cambridge: Harvard University Press.

CONNELL, J. H. & SOUSA, W. P., 1983. On the evidence needed to judge ecological stability or persistence. *American Naturalist, 121:* 789–824.

DAHLSTON, D. L. & ROWNEY, D. L., 1980. Influence of air pollution on population dynamics of forest insects and on tree mortality. In *Symposium on Effects of Air Pollution on Mediterranean Temperate Forest Ecosystems:* 125–131. Washington, U.S. Forestry Service (General Technical Report PSW-43).

DAYTON, P. K., 1975. Experimental evaluation of ecological dominance in a rocky intertidal algal community. *Ecological Monographs, 45:* 137–159.

DEN BOER, P. J., 1979. The significance of dispersal power for the survival of species, with special reference to the Carabid beetles in a cultivated countryside. *Fortschritte Zoologie, 25:* 79–94.

DEN BOER, P. J., 1981. On the survival of populations in a heterogeneous and variable environment. *Oecologia (Berlin), 50:* 39–53.

DIMBLEBY, G. W., 1978. Prehistoric man's impact on environments in Northwest Europe. In M. W. Holdgate & M. J. Woodman (Eds), *Breakdown and Restoration of Ecosystems:* 129–144. New York: Plenum Press.

DUNGAN, M. L., 1986. Three-way interactions: barnacles, limpets and algae in a Sonoran Desert rocky intertidal zone. *American Naturalist, 127:* 292–316.

FISHER, R. A., 1932. *The Design of Experiments.* Edinburgh: Oliver & Boyd.

FRANK, P. W., 1968. Life histories and community stability. *Ecology, 49:* 355–357.

GADGIL, M. & BOSSERT, W. H., 1970. The life-historical consequences of natural selection. *American Naturalist, 104:* 1–24.

GILBERT, J. M., 1959. Forest succession in the Florentine valley, Tasmania. *Papers of the Royal Society of Tasmania, 93:* 129–151.

GREEN, R. H., 1979. *Sampling Design and Statistical Methods for Environmental Biologists.* New York: Wiley.

GREEN, R. H. & HOBSON, K. D., 1970. Spatial and temporal structure in a temperate intertidal community with special emphasis on *Gemma gemma* (Pelecypoda: Mollusca). *Ecology, 51:* 999–1011.

HARPER, J. L., 1977. *The Population Biology of Plants.* London: Academic Press.

HARRIS, H. J., TALHELM, D. R., MAGNUSON, J. J. & FORBES, F. M., 1982. Green Bay in the future—a rehabilitative prospectus. *Great Lakes Fishery Commission Technical Report, 38.*

HEINSELMAN, M. L., 1973. Fire in virgin forests of the Boundary Waters Canoe Area, Minnesota. *Journal of Quaternary Research, 3:* 329–382.

HENDRICKS, A., HENLEY, D., WYATT, J. T., DICKSON, K. L. & SILVEY, J. K., 1974. Utilization of diversity indices in evaluating of a paper mill effluent on bottom fauna. *Hydrobiologia, 44:* 463–474.

HICKEY, J. J. & ANDERSON, D. W., 1968. Chlorinated hydrocarbons and eggshell changes in raptorial and fish-eating birds. *Science, 162:* 271–273.

HILBORN, R. & WALTERS, C. J., 1981. Pitfalls of environmental baseline and process studies. *Environmental Impact Assessment Review, 2:* 265–278.

HILL, J. & WIEGERT, R. G., 1980. Microcosms in ecological modelling. In J. P. Geici (Ed.), *Microcosms In Ecological Research.* Department of Energy, Symposium Series, 52: 138–163. Springfield: National Technical Information Service.

HOLLING, C. S., 1973. Resilience and stability of ecological systems. *Annual Review of Ecology and Systematics, 4:* 1–23.

HURLBERT, S. H., 1984. Pseudoreplication and the design of ecological field experiments. *Ecological Monographs, 54:* 187–211.

HUTCHINSON, G. E., 1951. Copepodology for the ornithologist. *Ecology, 32:* 571–577.

HUTCHINSON, T. C., 1973. Comparative studies of the toxicity of heavy metals to phytoplankton and their synergistic interactions. *Water Pollution Research of Canada, 8:* 68–90.

JERNAKOFF, P. & FAIRWEATHER, P. G., 1985. An experimental analysis of interactions among several intertidal organisms. *Journal of Experimental Marine Biology and Ecology, 94:* 71–88.

KENNELLY, S. J., 1987. Physical disturbances in an Australian kelp community. 1. Temporal effects. *Marine Ecology Progress Series, 40:* 145–153.

KEOUGH, M. J. & BUTLER, A. J., 1983. Temporal changes in species number in an assemblage of sessile marine invertebrates. *Journal of Biogeography, 10:* 317–330.

LEWIS, J. R., 1976. Long-term ecological surveillance: practical realities in the rocky littoral. *Oceanography and Marine Biology: Annual Review, 14:* 371–390.

LEWONTIN, R. C., 1969. The meaning of stability. *Brookhaven Symposium of Biology, 22:* 13–24.

LIVELY, C. M. & RAIMONDI, P. T., 1987. Desiccation, predation and mussel-barnacle interactions in the Northern Gulf of California. *Oecologia (Berlin), 74:* 304–309.

LOUCKS, O. L., 1970. Evolution of diversity, efficiency and community stability. *American Zoologist, 10:* 17–25.
MARGALEF, R., 1969. Diversity and stability: a practical proposal and a model for interdependence. *Brookhaven Symposium of Biology, 22:* 25–37.
McGUINNESS, K. A., 1988. *The Ecology of Botany Bay and the Effects of Man's Activities: a Critical Synthesis.* Sydney: Institute of Marine Ecology, University of Sydney.
MENGE, B. A., 1976. Organization of the New England rocky intertidal community: role of predation, competition and environmental heterogeneity. *Ecological Monographs, 46:* 355–393.
MENGE, B. A., 1978. Predation intensity in a rocky intertidal community. Relation between predator foraging activity and environmental harshness. *Oecologia (Berlin), 34:* 1–16.
MURDOCH, W. W., 1970. Population regulation and population inertia. *Ecology, 51:* 497–502.
NEWSOME, A. E., 1969. A population study of house-mice temporarily inhabiting a South Australian wheatfield. *Journal of Animal Ecology, 38:* 361–377.
NEWSOME, A. E. & CORBETT, L. K., 1975. Outbreaks of rodents in semi-arid and arid Australia: causes, preventions and evolutionary considerations. In I. Prakash & P. K. Ghosh (Eds), *Rodents in Desert Environments:* 117–153. The Hague: W. Junk.
ORIANS, G. H., 1974. Diversity, stability and maturity in natural ecosystems. In W. H. van Dobben & R. H. Lowe-McConnell (Eds), *Unifying Concepts in Ecology:* 139–150. The Hague: W. Junk.
PAINE, R. T., 1977. Controlled manipulations in the marine intertidal zone, and their contributions to ecological theory. *The Changing Scene in Natural Sciences, 1776–1976. Academy of Natural Science Special Publication, 12:* 245–270.
PETERSON, C. H. & BLACK, R., 1988. Density-dependent mortality caused by physical stress interacting with biotic history. *American Naturalist, 131:* 257–270.
PICKERING, A. D., 1981. Introduction: the concept of biological stress. In A. D. Pickering (Ed.), *Stress and Fish.* London: Academic Press.
PIELOU, E. C., 1974. *Population and Community Ecology.* New York: Gordon & Breach.
RAPPORT, D. J., REGIER, H. A. & HUTCHINSON, T. C., 1985. Ecosystem behaviour under stress. *American Naturalist, 125:* 617–640.
ROUGHGARDEN, J., IWASA, Y. & BAXTER, C., 1985. Demographic theory for an open marine population with space-limited recruitment. *Ecology, 66:* 54–67.
SCHELTEMA, R. S., 1971. Larval dispersal as a means of genetic exchange between geographically separated populations of shallow water benthic marine gastropods. *Bulletin of the Marine Laboratory, Woods Hole, 140:* 284–322.
SELYE, H., 1973. The evolution of a stress concept. *American Scientist, 61:* 692–699.
SIMBERLOFF, D. S., 1980. A succession of paradigms in ecology: essentialism to materialism and probabilism. In E. Saarinen (Ed.), *Conceptual Issues in Ecology:* 63–99. Dordrecht: Reidel Publishing.
SIMBERLOFF, D., 1986. The proximate causes of extinction. In D. M. Raup & D. Jablonski (Eds), *Patterns and Processes in the History of Life:* 259–276. Berlin: Springer-Verlag.
SNEDECOR, G. W. & COCHRAN, W. G., 1980. *Statistical Methods,* 7th edition. Iowa: Iowa State University Press.
SPIGHT, T. M., 1975. On a snail's chances of becoming a year old. *Oikos, 26:* 9–14.
STEWART-OATEN, A., MURDOCH, W. W. & PARKER, K. R., 1986. Environmental impact assessment: "pseudoreplication" in time? *Ecology, 67:* 929–940.
SUTHERLAND, J. P., 1974. Multiple stable points in natural communities. *American Naturalist, 108:* 859–873.
SUTHERLAND, J. P., 1981. The fouling community at Beaufort, North Carolina: a study in stability. *American Naturalist, 118:* 449–519.
SUTHERLAND, J. P. & ORTEGA, S., 1986. Competition conditional on recruitment and temporary escape from predators on a tropical rocky shore. *Journal of Experimental Marine Biology and Ecology, 95:* 155–166.
UNDERWOOD, A. J., 1974. On models for reproductive strategy in marine benthic invertebrates. *American Naturalist, 108:* 874–878.
UNDERWOOD, A. J., 1981. Techniques of analysis of variance in experimental marine biology and ecology. *Oceanography and Marine Biology: Annual Review, 19:* 513–605.
UNDERWOOD, A. J., 1985. Physical factors and biological interactions: the necessity and nature of ecological experiments. In P. G. Moore & R. Seed (Eds), *The Ecology of Rocky Coasts:* 372–390. London: Hodder & Stoughton.
UNDERWOOD, A. J., 1986a. What is a community? In D. M. Raup & D. Jablonski (Eds), *Patterns and Processes in the History of Life:* 351–367. Berlin: Springer-Verlag.
UNDERWOOD, A. J., 1986b. The analysis of competition by field experiments. In D. J. Anderson & J. Kikkawa (Eds), *Community Ecology: Pattern and Process:* 240–260. Melbourne: Blackwell Scientific Press.
UNDERWOOD, A. J. & PETERSON, C. H., 1988. Towards an ecological framework for investigating pollution. *Marine Ecology Progress Series, 46:* 227–234.
UNDERWOOD, A. J., DENLEY, E. J. & MORAN, M. J., 1983. Experimental analyses of the structure and dynamics of mid-shore rocky intertidal communities in New South Wales. *Oecologia (Berlin), 56:* 202–219.
WESTMAN, W. E., 1978. Measuring the inertia and resilience of ecosystems. *BioScience, 28:* 705–710.

WESTMAN, W. E. & O'LEARY, J. F., 1986. Measures of resilience: the response of coastal sage scrub to fire. *Vegetatio, 65:* 179–189.

WHITBY, L. M., STOKES, P. M., HUTCHINSON, T. C. & MYSLIK, G., 1976. Ecological consequence of acidic and heavy-metal discharges from the Sudbury Smelters. *Canadian Mineralogy, 14:* 47–57.

WHITE, T. C. R., 1974. A hypothesis to explain outbreaks of looper caterpillars, with special reference to populations of *Selidosema suavis* in a plantation of *Pinus radiata* in New Zealand. *Oecologia (Berlin), 16:* 279–301.

WHITE, T. C. R., 1976. Weather, food and plagues of locusts. *Oecologia (Berlin), 22:* 119–134.

WHITE, T. C. R., 1978. The importance of a relative shortage of food in animal ecology. *Oecologia (Berlin), 33:* 71–86.

Biological Journal of the Linnean Society (1989), *37:* 79–99. With 5 figures

Structured population models: a tool for linking effects at individual and population level

R. M. NISBET, W. S. C. GURNEY

Department of Physics and Applied Physics, University of Strathclyde, Glasgow G4 0NG

W. W. MURDOCH

Department of Biological Sciences, University of California, Santa Barbara, CA 93106, U.S.A.

AND

E. McCAULEY

Department of Biology, University of Calgary, Calgary, Alberta, T2N 1N4, Canada

We address the problem of relating information on the effects of a particular stress on *individuals* to possible effects at the *population* level. Structured population models aim to predict population dynamics from a careful specification of the dynamics of individuals; however, in spite of major mathematical advances, there are only a few cases where such models have made significant contributions to ecological understanding. This paper reports progress to date on a project in which we construct both individual and population models of *Daphnia*. We present a model of individual growth and development which has been tested against results from several laboratories on *D. pulex*. We propose a simple, stage-structured population model and give a preliminary report of some of its properties.

KEY WORDS:—Structured population model – population dynamics – *Daphnia* energy allocation – population cycles.

CONTENTS

0024–4066/89/050079+21 $03.00/0

INTRODUCTION

One fundamental impediment to elucidating the concept of stress in ecology is that the simplest measurements to undertake are often on *individuals*, while our primary interest is likely to be in effects at the *population* level. For example, the direct effects of toxic substances may be to inhibit growth or development, reduce fecundity or increase mortality; the consequent changes in population densities and in the pattern of population fluctuations depend on the regulatory mechanisms of the population under investigation. It follows that a potentially valuable tool in the study of ecological stress will be *structured population models* which aim to predict the dynamics of a population, given a well-posed, dynamic specification of the response of individual members of that population to external factors.

Systematic mathematical methods for the formulation and analysis of structured population models have been developed in recent years (e.g. Metz & Diekmann, 1986), but there has been much less progress in the equally demanding task of assessing their practical utility. The appropriate level of complexity of a model for a particular task remains a matter of considerable controversy, progress towards whose resolution would be considerably assisted by some detailed case studies. We therefore have started a programme in which we plan to contrast the performance of simple and complex models of the zooplankter *Daphnia*. These animals are particularly appropriate for this work as there exists a vast body of literature on the physiology of stressed and unstressed individuals as well as data on laboratory and natural populations.

Many natural *Daphnia* populations exist at low food levels, sustained by a balance between low fecundity and (presumably) mortality. Two of us (Murdoch & McCauley, 1985; McCauley & Murdoch, 1987; McCauley, Murdoch & Watson, 1988) have previously conjectured that the diverse patterns of *Daphnia* dynamics that follow a spring algal peak reflect the *Daphnia*–food interaction rather than exogenous forcing by biotic or abiotic factors. If these conjectures are valid, it follows that any model of *Daphnia* population dynamics must treat carefully the assimilation and utilization of food.

McCauley & Murdoch (1987) also showed that quasi-cyclic fluctuations in both laboratory and field populations of *Daphnia* have a dominant period close to the generation time and (where demographic data are available) that the fluctuations have the following pattern: a burst of reproduction at low density producing a peak population consisting largely of juveniles, then a long period of declining density, suppressed reproduction, and slow juvenile development, the population at its nadir consisting largely of adults which eventually produce the next burst of recruits. These observations support our premise that a realistic population model must incorporate some aspects of the physiological structure (e.g. proportions of large/small, old/young individuals) of the population.

As already noted, development of such a structured population model starts with the construction of a model of the properties of individuals, and in previous papers (McCauley *et al.*, 1989b; Gurney *et al.*, 1989) we developed one such model for *Daphnia*. For a detailed exposition we refer the reader to Gurney *et al.* (1989), but its main assumptions are set out in section 2 of the present paper. The emphasis of that section is on the judgements that were necessary to construct the model and not on the technical details. In similar spirit we refer the reader to the original paper for detail of our quantitative tests of the model, but highlight certain qualitative predictions on starvation and recovery, an area where the model suggests new critical experiments.

In section 3, we discuss the problems inherent in constructing a population model based on our rather elaborate description of individual physiology, and we expose certain formidable technical obstacles that preclude immediate incorporation of our model within the standard mathematical framework for structured population models. To circumvent these problems, which are the subject of current research, we simplify our description of individual physiology, so as to permit development of a model in which the population dynamics are described in terms of a set of coupled delay-differential equations. Quantitative tests of the model are still in progress, but we discuss its ability to explain the demography of the cycles in real *Daphnia* populations.

A MODEL OF GROWTH AND REPRODUCTION IN INDIVIDUAL DAPHNIA

The model developed in our two previous papers (McCauley *et al.*, 1989b; Gurney *et al.*, 1989) describes dynamically the utilization of food by daphnids for growth, maintenance and reproduction; for this we need a description of 'food'. Notwithstanding evidence that the composition of the algal population, and in particular the edible fraction, may significantly influence *Daphnia* population dynamics (McCauley *et al.*, 1988 and references therein) we initially regard food as a homogeneous assemblage within the water, describable by a single density, namely carbon content per unit volume. The model describes the fate of this carbon following ingestion, and specifies a set of 'sinks' or 'pools' within a daphnid, together with a set of rules for allocating assimilate to these pools. The model equations are summarized in Table 1.

(A) The model structure and equations

Reserves and the short-term fate of assimilate

There is good evidence (e.g. radioisotope studies of Lampert (1975)) that assimilate is incorporated into the body structure within a few hours, so we do not need a representation of short-term reserves in the model. There is also good evidence from starvation experiments that as much as 70% of body tissue may be used as long-term reserves during periods of starvation. We therefore assume that (long-term) reserves constitute a specified fraction of the normal body weight (excluding eggs) of a well-fed animal; an animal dies of starvation if its body weight drops below the appropriate fraction of the 'normal' weight (defined later).

In our new energy channelling scheme (Fig. 1), we assume that assimilate is committed *immediately and irreversibly* to reproduction, or to growth and maintenance. We then assume that given 'sufficient' food, an animal of length L

TABLE 1. The model of individual growth and reproduction as specified by Gurney *et al.* (1988). The equations are set out in a style close to that required for computer implementation of the model. The symbol ← means 'is assigned the value'

Food assimilation and utilization	$I_{max} = SL^Q[1 - \exp\{-(L/L_u)^i\}]$
	$I = I_{max}F/(F + F_h)$
	$A = \varepsilon_A I$
	$A_{max} = \varepsilon_A I_{max}$
	If $L < L_m$ then $K_u = 1.0$
	else $K_u = K_{min}(A_K + 1)/(A_K + A/A_{max})$
	If $K_u < 1.0$ then $K_o(L, A) = K_u + (1 - K_u)\exp\{(L_m - L)/L_s\}$
	else $K_o(L, A) = 1.0$
At moult	$E_{max} = T_m[1 - K_o(L, A_{max})]\varepsilon_A I_{max}$
	$W_e = W_{eo} + W_{es}E/E_{max}$
	If $E > W_e$ then $\{N_{ed} \leftarrow E/W_e, E \leftarrow 0\}$
	else $\{N_e \leftarrow 0$, E unchanged$\}$
	If $W > (L/X)^P$ then $\{L \leftarrow XW^{1/P}\}$
	else {L unchanged}
Between moults	$M = BW^Y$
	If $A < M$ or $W < (L/X)^P$ then $K = 1.0$
	else $K = K_o(L, A)$
	$dN_e/dt = dL/dt = 0$
	$dE/dt = (1 - K)A$
	$dW/dt = KA - M$
Food dynamics	Semichemostat: $dF/dt = D(F_r - F) - I/V$
	Transfer $dF/dt = -I/V$ between transfers
	$F \leftarrow F_r$ at transfers

allocates a constant proportion of assimilate to reproduction and the remaining fraction to growth and maintenance combined.

The utilization of assimilate when food is scarce

We need rules for energy channelling when food supply varies as well as when it is constant. Following Kooijman (1986a, but noting that he was modelling commitment from a reserve pool and not the immediate fate of assimilate), a natural assumption is that if the default allocation of assimilate to growth and

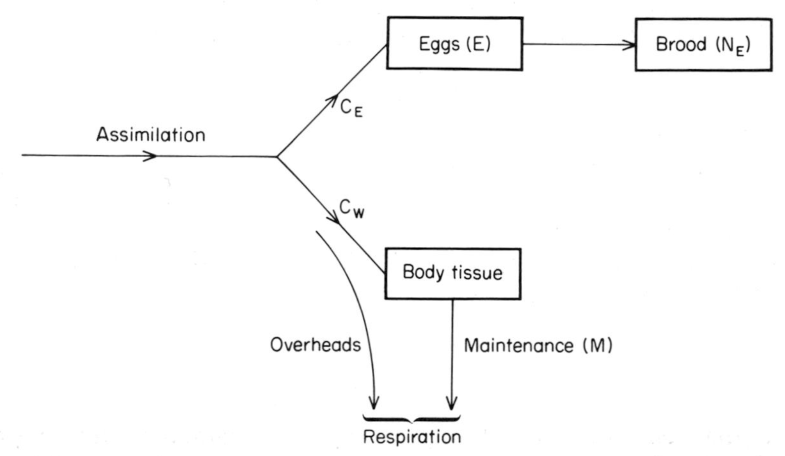

Figure 1. The energy allocation scheme in the model of individual growth and reproduction (from McCauley *et al.*, 1989b).

maintenance is at any time insufficient to meet maintenance, then a daphnid meets immediate maintenance needs by (in order): (i) stopping growth, (ii) supressing commitment to new eggs, (iii) metabolizing reserves. However, although qualitatively plausible, a model embodying the strict priorities outlined above made predictions inconsistent with experiments on the growth of individuals in continuous flow systems, and it appears that at moderately low food densities daphnids give slightly higher priority to growth than is implied by the above rules (Gurney *et al.*, 1989).

In addition, *Daphnia* is able to recover from periods at low food levels and eventually perform at a higher food level at the same rate as individuals raised continuously at that higher level (Ingle, Wood & Banta, 1937; Kooijman, 1986a). We therefore hypothesize that starving individuals give priority to growth over reproduction whenever their weight is less than a notional 'weight-for-length'. We assume that this weight-for-length is the weight of the body (but not the eggs) of a healthy daphnid of length L immediately after the moult, there being experimental support for the existence of a simple allometric relationship between these quantities. We are thus introducing a practical criterion to define a starving animal, and then postulating that starving animals suppress reproduction in favour of recovering body weight (i.e. reserves). If the appropriate weight-for-length is reached, the normal allocation rules apply.

The rules for energy channelling

We now define the variables for our model and specify the set of mathematical rules governing the utilization of assimilate by the animal.

We recognize that *Daphnia* development proceeds through a series of discrete instars separated by moults. We assume that instar duration is a constant (T_m) for all instars and at all food levels (though both of these assumptions can be relaxed without prejudice to the structure of the model).

We characterize a daphnid of age a by the following five *state variables:* carapace length (L), body weight excluding eggs and material destined for eggs (W), weight-for-length (W_{fl}) as introduced in the previous subsection, material in the body committed to egg production (E), and number of eggs in the brood pouch (N_E). We assume (section B) that for well-fed daphnids, weight and length are related immediately after a moult by an allometric relation of the form

$$W = (L/X)^p, \tag{1}$$

where X and p are constants.

Assimilate is assumed to be allocated between growth, reproduction and maintenance in accordance with the scheme shown in Fig. 1, where we also introduce notation for the various material fluxes. With this partitioning, elementary book-keeping yields the following differential equations which are assumed to hold *throughout an intermoult*:

$$dW/dt = C_w - M, \tag{2}$$

$$dE/dt = C_E. \tag{3}$$

We assume that length cannot change during an intermoult, implying a similar property for W_{fl}. Since eggs are only released at a moult, N_E cannot change during an intermoult.

At a moult, all the state variables except W take new values as follows:

$$L = XW^{1/p}, \tag{4}$$

unless this would produce a *decrease* in length, in which case the value of L is unchanged. The new weight-for-length is then calculated from the new length as

$$W_{fl} = (L/X)^p. \tag{5}$$

All assimilate committed to reproduction during the previous instar is passed as new eggs to the brood pouch (the existing clutch being released as neonates). We assume that a certain quantity of assimilate (W_e), whose value depends on the amount of material available for egg production, is required to produce an egg; thus the new brood size is given by

$$N_E = \begin{cases} E/W_e \text{ if } E > W_e \\ 0 \text{ otherwise} \end{cases} \tag{6}$$

where

$$W_e = W_{eo} + W_{es}E/E_{max}. \tag{7}$$

For simplicity, present implementations of the model allow fractional eggs to be passed to the brood pouch, but in principle equation (6) could be replaced by a more elaborate rule in which an integer number of eggs are passed to the brood pouch with the remaining egg material retained in the body for the next instar (as is done if there is insufficient material for even one egg). However, given the large observed variability in brood size, even for *Daphnia* grown under controlled conditions, such elaboration would appear rather unnecessary—hence the simple form for equation (6). With this simple form, the animal normally starts the next moult with no carry-over of egg material; at the start of each instar we set $E = 0$, unless the brood pouch is empty ($N_e = 0$) in which case we leave E unchanged.

Dynamically, the key feature of the model is the 'K-switch' in Fig. 1 which determines the fluxes C_E and C_w. We define this quantity K (which is the fraction of assimilate that goes to growth plus maintenance, leaving the fraction $1-K$ for reproduction) by writing the fluxes formally in the form

$$C_w = KA, \qquad C_E = (1-K)A, \tag{8}$$

(where A is the rate of assimilating food), so the specification of the model *structure* is complete once we select the rule for calculating the quantity K. We have already noted that this allocation function must depend on length (to distinguish juveniles from adults), assimilation rate (to meet maintenance when food is scarce), and weight-for-length (to give priority to growth in animals recovering from starvation). After considering a variety of forms (Gurney *et al.*, 1989), we concluded that the structure of these dependences was given by the rule

$$\begin{aligned} &\text{If } A < M \text{ or } W < (L/X)^p \text{ Then } K = 1.0 \\ &\qquad\qquad\qquad\qquad\qquad \text{Else } K = K_o(L, A). \end{aligned} \tag{9}$$

The form of the function K_o is given in the next section (equation 16).

Implementing the model now requires us to specify functional forms for the equations relating assimilation (or ingestion) to length, maintenance to weight, and the default partitioning function $K_o(L, A)$ to length and assimilation rate. This is the subject of the next section. However, it is important to realizc that the energy-channelling scheme of the above section does not rely on these particular forms. Thus when testing the model against experimental information it is important to distinguish *a priori* whether we are testing the fundamental partitioning rules, or the effects of any particular functional form or parameter selection.

(B) The model functions and parameters

Length and weight

The model as formulated in the previous section requires a relationship between length and weight (in carbon units) for a daphnid stripped of eggs immediately after the moult. The literature contains an apparently wide range of observed relationships between length and carbon content or dry weight, but careful cxamination of these results for one species (*D. pulex*) shows that most of the data are consistent with the interpretation that there is a single allometric relationship between carbon content and carapace length, valid for both adults and juveniles, which is, at most, weakly dependent on the food regime in which the animals were living. This is given in equation (1) above.

Ingestion and assimilation

For daphnids of a specified length, the dependence of feeding rate (I) on food density (F) can adequately be described by a (type 2) hyperbolic functional response; thus ingestion rate can be written in the form

$$I = I_{max}F/(F+F_h). \tag{10}$$

Our review of existing data (McCauley *et al.*, 1989b; Gurney *et al.*, 1989) suggests that we may safely assume that the half-saturation constant (F_h) does not vary with length.

McCauley *et al.* (1989b) discussed the variation with length in the maximum ingestion rate (I_{max}), which is well fitted by the function

$$I_{max} = SL^Q\{1-\exp(-(L/L_u)^i)\}. \tag{11}$$

We follow Lampert (1975) and regard food as *assimilated* if it passes across the wall of the gut, the assimilation efficiency then being defined as the ratio of instantaneous assimilation rate to ingestion rate. Measurements of this quantity exhibit high variability, but it is not established whether the variation is systematically related to either food concentration or the size of the individual. Consequently, for our model we assume a constant assimilation efficiency, assimilation rate A then being given by

$$A = \varepsilon_A I. \tag{12}$$

Maintenance

This is possibly the most elusive quantity on which we require high quality information for the model. We argued in McCauley *et al.* (1989b) that maintenance rate could be modelled as the sum of two components, the first a term proportional to weight, the second a representation of the continuous commitment of new material to make the next carapace. We thus set

$$M = \beta W + T_m^{-1} W_s, \quad (13)$$

where T_m is the intermoult duration, and W_s is the weight of a cast skin which varies with body weight as

$$W_s = 0{\cdot}016\, W^{1.47} \quad (14)$$

when both quantites are measured in units of micrograms dry weight (Lynch, Weider & Lampert, 1986). It turns out that the use of equations (13) and (14) to specify maintenance makes simple analytic calculations of such quantities as maximum length or brood size unacceptably awkward. For practical convenience we therefore assume an allometric maintenance-weight relationship of the form

$$M = BW^Y \quad (15)$$

and calculate parameters B and Y so as to make the resulting curve as close as possible (which turns out to mean almost indistinguishable to graphical accuracy) from that implied by equations (13) and (14).

For detailed justification of these assumptions we refer the reader to McCauley *et al.* (1989b). However, it is important to note that our representation, in which total *maintenance* scales as weight to some power greater than one, is consistent with the common claim (e.g. Peters, 1983) that *respiration* in ectotherms scales as $W^{0.75}$. The reconciliation comes through work of Kooijman (1986b) who recently interpreted many observed relationships between respiration rate and weight as the sum of two terms: the true routine metabolism (assumed proportional to weight) and a term representing the overheads on growth (assumed proportional to the instantaneous assimilation rate).

Allocation to reproduction

Gurney *et al.* (1989) propose that the dependence of the allocation function on assimilation rate is treated by first introducing a quantity K_u:

$$\begin{aligned} &\text{If } L < L_m \text{ then } K_u = 1.0 \\ &\qquad\qquad\quad \text{else } K_u = (A_k + 1)/(A_k + A/A_{max}). \end{aligned} \quad (16a)$$

McCauley *et al.* (1989b) give a table, based on data in Paloheimo, Crabtree & Taylor (1982), from which we can infer the length dependence of the allocation function $K_o(L, A)$, a reasonable representation of very sparse data being obtained by fitting a clipped exponential function. Finally, K_o is calculated from the rule

$$\begin{aligned} &\text{If } K_u < 1.0 \text{ then } K_o(L, A) = K_u + (1 - K_u) \exp\{(L_m - L)/L_s)\} \\ &\qquad\qquad\quad \text{else } K_o(L, A) = 1.0 \end{aligned} \quad (16b)$$

in which the exponential term represents a gradual 'switch-on' of commitment to reproduction before and after the primiparous instar.

TABLE 2. The parameter set for the model of individual growth and reproduction for *D. pulex* at 20°C–from Gurney *et al.* (1989)

Parameter	Value	Units	Brief description
S	9.64×10^{-3}	$mgC\ day^{-1}\ mm^{-Q}$	Constant in ingestion function
Q	1.76	dimensionless	Index in ingestion function
L_u	0.95	mm	Constant in ingestion function
i	2.14	dimensionless	Index in ingestion function
F_h	0·164	$mgC\ l^{-i}$	Half saturation constant
ε_A	0.6	dimensionless	Assimilation efficiency
B	0.28	$(mgC)^{1-Y} day^{-1}$	Coefficient in maintenance —weight relationship
Y	1.14	dimensionless	Allometric index in maintenance —weight relationship
K_{min}	0.18	dimensionless	Minimum allocation to growth + maintenance
A_K	0.15	dimensionless	Parameter in allocation function
L_m	0.9	mm	Minimum length to allocate energy to reproduction
L_s	0.33	mm	Constant in energy partitioning formula
P	2.4	dimensionless	Allometric index for weight-length conversion
W_{eo}	0.4×10^{-3}	$mgC\ egg^{-1}$	Minimum weight of an egg
W_{es}	$1{\cdot}4 \times 10^{-3}$	$mgC\ egg^{-1}$	Constant in formula determining egg weight
L_n	0.6	mm	Length of a neonate
T_m	2	day	Average instar duration

The model parameters

We have derived one complete set of parameters for the model—for *D. pulex* at 20°C. These are presented in Table 2. Preparation of this table involved making a number of judgements additional to those required in formulating the model, for example selecting between conflicting data on ingestion and assimilation rates. We refer the reader to the appendix of Gurney *et al.* (1989) for further details.

(C) Testing the model

Qualitative predictions on starvation and recovery

From the start of this paper we have recognized that the ability to model life at low food is vital; indeed the fundamental structure of the model, and much of its complexity, is the result of attention to detail in this regard. Thus before proceeding to quantitative tests of the model, it is appropriate to look qualitatively at its predictions regarding starvation and recovery.

If an animal is assumed to be capable of surviving a bout of starvation until its weight drops to a specified fraction (f) of its weight-for-length, then it is straightforward to derive an approximation to the starvation time (T_s) of an individual, namely

$$T_s = \log_e f/(\text{fractional daily maintenance rate at start of starvation}). \quad (17)$$

Since we relied heavily on qualitative information from starvation experiments in formulating the model, and used data on weight loss during starvation as part of our calculation of the maintenance parameters, prediction of starvation times for *D. pulex* is not a valid test of the model. However, the model also makes predictions about recovery from starvation, which ought to be amenable to experimental test. Suppose an adult has starved for a few days and has a weight less than its weight for length. If it is now introduced to food of constant density, our energy allocation rules imply that it allocates all assimilate to growth and maintenance until it regains its weight-for-length. It can be shown (after some algebraic manipulation) that there are now three possibilities.

(i) If the food density is sufficiently high, the weight-for-length is eventually achieved and commitment to reproduction resumes.

(ii) At sufficiently low food density, weight continues to decline until it reaches the value f times the weight-for-length, whereupon the animal dies of starvation.

(iii) For an intermediate range of food levels, the weight grows or declines to an asymptotic level where the animal neither dies of starvation nor resumes reproduction.

We know of no existing data against which these predictions may be tested, but have in progress a series of experiments on starvation/recovery designed to test them.

Quantitative predictions on growth and reproduction

The effort involved in constructing a model as elaborate as this is only justifiable if it results in quantitative predictions on growth and reproduction, the latter in particular being vital for a population model. In Gurney *et al.* (1989) we reported a series of tests of the model against data from four different investigators (detailed in Table 3), using different clones of *D. pulex* and different experimental procedures. All but one of the sets of experiments (Taylor, 1985) used a *transfer culture* technique in which animals were grown in individual containers and transferred at regular intervals to new containers with fresh food. We selected four quantities to characterize the growth and reproduction of individuals at each food density: the maximum observed length, maximum brood size, length at first instar containing eggs, and time to first placing of eggs in the brood pouch. In Fig. 2 we show a comparison of predicted and observed values of these quantities; all except length at first brood (a quantity in which there is no obvious pattern to the original data) are well predicted by the model.

TABLE 3. Data sets used to test the model of *Daphnia* growth and development. From Gurney *et al.* (1989): A = *Chlamydomonas reinhardii*; B = *Scenedesmus acutus*

Worker(s)	Number of individuals per container	Food type	Food conclusion ($mgC\ l^{-1}$)	Value (ml)	Renewal
Richman (1958)	1	A	0.58	10	Daily transfers
	1	A	1.15	10	Daily transfers
	1	A	1.73	10	Daily transfers
	1	A	2.30	10	Daily transfers
Paloheimo *et al.*, (1982)	1	A	4.2	100	Daily transfers
Taylor (1985)	5–15	B	0.1	200	Dilution ($7.5\ d^{-1}$)
Taylor & Gabriel (1985)	5–15	B	0.2	200	Dilution ($7.5\ d^{-1}$)
	5–15	B	1.0	200	Dilution ($7.5\ d^{-1}$)
Lynch *et al.*, (1986)	1	A+B	1.54	40	Two-day transfers

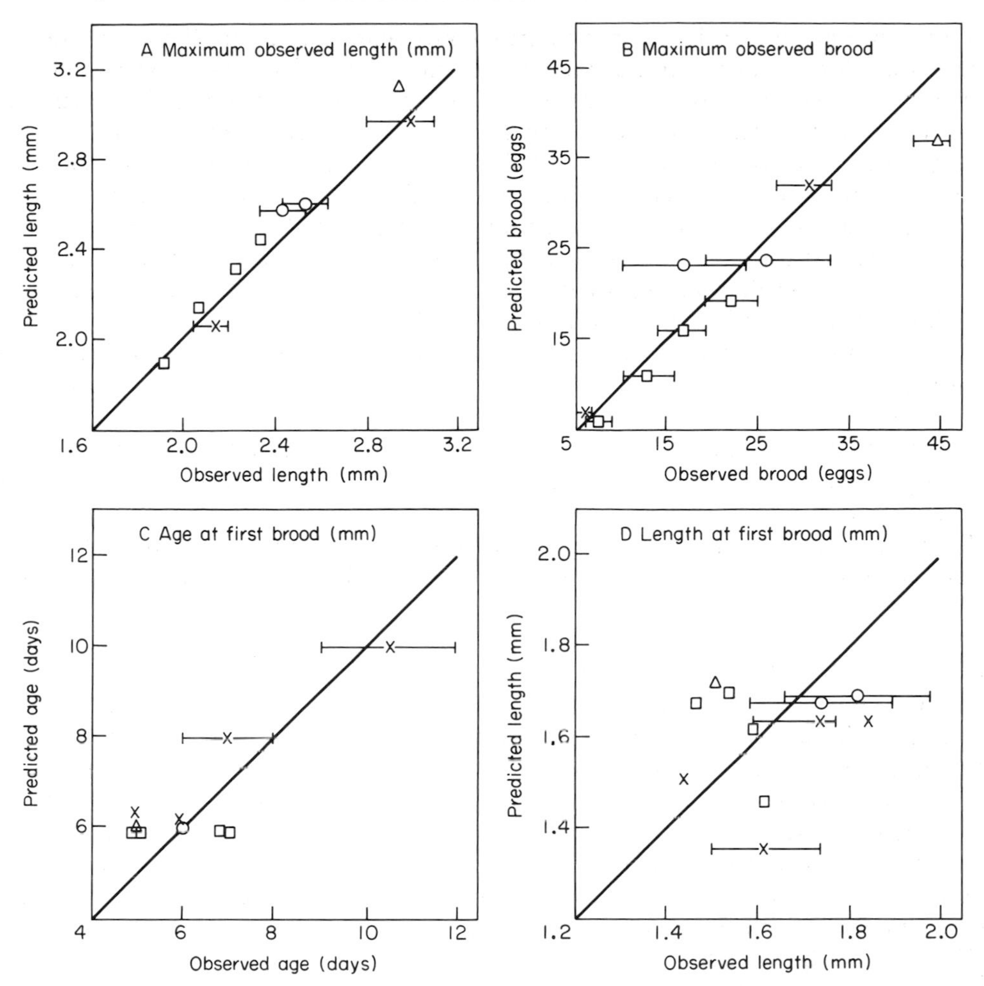

Figure 2. Predicted and observed life-history characterizations for the model of individual growth and reproduction of *D. pulex*. The sources of data (detailed in Table 3) are: Paloheimo *et al.*, 1982 (□), Lynch *et al.*, 1986 (○), Richman, 1958 (△), and Taylor, 1985 (×). Figure reproduced from Gurney *et al.* (1989).

This success is an encouraging start to the wider programme of relating individual and population phenomena; at least it is possible to construct a model in which parameters derived largely from short-term measurements of physiological rate processes are used to successfully predict individual growth and reproduction.

A STAGE-STRUCTURED POPULATION MODEL

(A) Model formulation

Having constructed a model of individuals we might now expect to be able to use techniques similar to those in Metz & Diekmann (1986: chapter 3) to construct a population model. Their approach, which to the best of our understanding reflects the current mathematical state of the art, involves selecting a set of variables which specify the *state* of an individual (i-state); the individual model then gives us the differential equations which describe the change with time of these state variables. The population dynamics are given by partial differential equations or integral equations which are numerically tractable only if the equations in the individual model are reasonably well behaved. Unfortunately, certain key variables in our individual model change discontinuously (at moults), while derivatives of all the state variables may change discontinuously, for example when food drops so that an animal enters starvation. We are therefore some way away from being able to follow this approach to a population model.

The same discontinuities cause difficulty if we adopt a 'brute force' approach and, arguing that a population is simply a collection of individuals, model a small volume containing (say) tens of individuals by solving numerically the differential equations for each individual. There are now many non-analytic points in the solutions, and it is our experience that these have to be located rather accurately to avoid large numerical error. These problems are superable (but only with considerable effort) and we are pursuing this approach further.

Faced with these difficulties, we have temporarily compromised our original objective of deriving a structured model based strictly on the individual model, and formulated a stage-structured 'continuous development' model in which the life history is subdivided into a number of physiological stages, within which development and mortality rates vary continuously and are assumed to have the same values at any given time. The methodology has been described in detail elsewhere (Gurney, Nisbet & Lawson, 1983; Nisbet & Gurney, 1983; Gurney, Nisbet & Blythe, 1986), here we merely note that the pay-off from this simplification is that the population dynamics may be described in terms of a relatively small number of delay-differential equations whose numerical solution is fairly straightforward.

In our stage-structured model, we recognize three stages—juveniles (whose density is denoted by J), young adults (density Y), and mature adults (density A)—all of which are assumed to eat a common food of (carbon) density F. *Juveniles* are defined to be individuals which commit, or have the capacity to commit, all excess assimilate not required for maintenance, to growth. The *young adult* stage covers the period (roughly two moults) that elapses between the first commitment (or capacity for commitment) of assimilate to egg production and the release of the first brood of neonates. The *adult* stage covers the entire remainder of the animal's life. The densities for the three stages then must satisfy the following balance equations

$$dJ(t)/dt = R_J(t) - M_J(t) - m_J(t)J(t) \tag{18}$$

$$dY(t)/dt = R_Y(t) - M_Y(t) - m_A(t)Y(t) \tag{19}$$

$$dA(t)/dt = R_A(t) - m_A(t)A(t) \tag{20}$$

which are coupled through the various vital rate functions to an equation describing the balance of supply and consumption of food namely

$$dF(t)/dt = R_F(t) - J(t)I_J(t) - [Y(t) + A(t)]I_A(t). \tag{21}$$

In these equations, at time t: $R_F(t)$ = food replacement rate per unit volume;

$I_J(t)$ = feeding rate (mgC per day) for juveniles; $I_A(t)$ = feeding rate (mgC per day) for adults; $R_J(t)$ = recruitment rate per unit volume of juveniles; $R_Y(t)$ = recruitment rate per unit volume of young adults; $R_A(t)$ = recruitment rate per unit volume of adults; $M_J(t)$ = maturation rate per unit volume of juveniles; $M_Y(t)$ = maturation rate per unit volume of young adults; $m_J(t)$ = per capita death rate of juveniles; $m_A(t)$ = per capita death rate of adults.

For each stage we are required to define a *development index* (or physiological age) such that individuals mature to the next stage when their development index attains a particular value. This permits us to relate the rate of maturation out of a stage to the rate of recruitment to the stage at some previous time. The remaining steps in the formulation of the model involve assumptions on mortality within and between stages, and assumptions on fecundity. A full account of the basis for these assumptions and of the details of parameter estimation will be published elsewhere; however as with the individual model, following the programme through involves making a significant number of biological judgements and using our own data to fill gaps in the literature. These aspects we highlight in the next two subsections (B and C), and some preliminary tests of the model are presented in subsection D. The full set of equations defining the model is given in Table 4 and a provisional set of parameter values in Table 5.

TABLE 4. Formulae for the 'continuous development' *Daphnia* population model

Food and feeding:	$F(t) = R_F(t) - J(t)I_J(t) - [Y(t) + A(t)]I_A(t)$ $I_j = \frac{I_{mj}F}{F+F_h} \quad I_A = \frac{I_{ma}F}{F+F_h}$ $\bar{I}_J(t) = T_m^{-1} \int_{t-T_m}^{t} I_J(x)\,dx \quad \bar{I}_A(t) = T_m^{-1} \int_{t-T_m}^{t} I_A(x)\,dx$
Juvenile development:	$h_J(t) = T_{max}^{-1} + [T_{min}^{-1} - T_{max}^{-1}] \frac{[\varepsilon_{Aj} I_J(t) - \Gamma_J]_+}{[\varepsilon_{Aj} I_{mj} - \Gamma_J]}$
Fecundity:	$\beta(t) = e[\varepsilon_{Aa}\bar{I}_A(t - T_m) - \Gamma_A]_+$
Mortality:	$m_J(t) = m_{oj} + m_{sj} \exp[-\bar{I}_J(t)/I_{JO}]$ $m_A(t) = m_{oa} + m_{sa} \exp[-\bar{I}_A(t)/I_{AO}]$
Juvenile development times:	$\dot{\tau}_J(t) = 1 - h_J(t)/h_J(t - \tau_J(t))$
Through-stage survival:	$\dot{P}_J(t) = P_J(t)\left[m_J(t - \tau_J(t)) \frac{h_J(t)}{h_J(t - \tau_J(t))} - m_J(t)\right]$ $\dot{P}_Y(t) = P_Y(t)[m_A(t - 2T_m) - m_A(t)]$ $S_{JY}(t) = 1 - \exp[-(T_{max} - T_J(t)/T_o)]$
Recruitment and maturation:	$R_J(t) = \beta(t)A(t)$ $M_J(t) = R_J(t - \tau_J(t))P_J(t) \frac{h_J(t)}{h_J(t - \tau_J(t))}$ $R_Y(t) = M_J(t)S_{JY}(t)$ $M_Y(t) = R_Y(t - 2T_m)P_Y(t)$ $R_A(t) = M_Y(t)$
Population balance equations	$\dot{J}(t) = R_J(t) - M_J(t) - m_J(t)J(t)$ $\dot{Y}(t) = R_Y(t) - M_Y(t) - m_A(t)Y(t)$ $\dot{A}(t) = R_A(t) - m_A(t)A(t)$

TABLE 5. A provisional parameter set for the 'continuous development' population model—appropriate to *D. pulex* at 20°3

Parameter	Value	Units	Brief description
T_{max}	20	day	Maximum development time: juveniles
T_{min}	4	day	Minimum development time: juveniles
T_o	3.5	day	Constant in formula for surviving maturation
e	1000	$(mgC)^{-1}$	Conversion efficiency—food to neonates
F_h	0.164	$mgC\ l^{-1}$	Half saturation constant
I_{mj}	6.5×10^{-3}	$mgC\ day^{-1}$	Maximum feeding rate: juveniles
I_{ma}	2.1×10^{-2}	$mgC\ day^{-1}$	Maximum feeding rate: adults
Γ_J	3.2×10^{-4}	$mgC\ day^{-1}$	Maintenance rate: juveniles
Γ_A	1.1×10^{-3}	$mgC\ day^{-1}$	Maintenance rate: adults
ε_{Aj}	0.6	dimensionless	Assimilation efficiency: juveniles
ε_{Aa}	0.6	dimensionless	Assimilation efficiency: adults
T_m	3.0	day	Average instar duration
I_{oj}	4×10^{-4}	$mgC\ day^{-1}$	Constant in formula for juvenile mortality
I_{oa}	2×10^{-3}	$mgC\ day^{-1}$	Constant in formula for adult mortality
m_{oj}	3×10^{-3}	day^{-1}	Background juvenile mortality
m_{oa}	3×10^{-3}	day^{-1}	Background adult mortality
m_{sj}	0.217	day^{-1}	Maximum mortality increment: juveniles
m_{sa}	0.217	day^{-1}	Maximum mortality increment: adults

(B) Development and maturation of juvenile daphnids

The primary role of the juvenile development index in the population model is to determine the timing of the onset of maturation of the young adult stage. From Fig. 2D and from our own experiments, it is clear that over a wide range of food densities, the first brood appears at a certain critical length. However, our own experiments (McCauley, Murdoch and Nisbet, 1989b) point in addition to the existence of an upper duration to the juvenile stage; *D. pulex* grown at very low food densities at 20°C which survive to an age of around 20 days will attempt to produce an egg, sometimes dying as a result. Taken together, these observations point to a development index, q, which is a weighted average of age and weight. If we chose to make the development index dimensionless and specify that neonates have $q = 0$, and that maturation occurs at $q = 1$, and further accept a technical constraint that an individual's development index must never decrease as she grows older, then this suggests a development rate of the form

$$h(t) = dq/dt = T_{max}^{-1} + \frac{T_{min}^{-1} - T_{max}^{-1}}{[\varepsilon_{Aj} I_{mj} - \Gamma_j]} [\varepsilon_{Aj} I_j(t) - \Gamma_j]_+ \tag{22}$$

where T_{max} and T_{min} are respectively the minimum and maximum development times for a juvenile, ε_{Aj} represents the assimilation efficiency of a juvenile, Γ_j its maintenance requirements ($mgC\ day^{-1}$), and I_{mj} its maximum possible feeding rate. The notation $[\ ]_+$ is a shorthand form for the rule that if the value of the quantity in square brackets is negative, it is replaced by zero when evaluating the formula.

Figure 3 shows that, with this choice of development index, there is a satisfactory fit to both our own measurements of juvenile stage duration and to the data used by Gurney *et al.* (1989) and reproduced in Fig. 2 in the tests of our individual model.

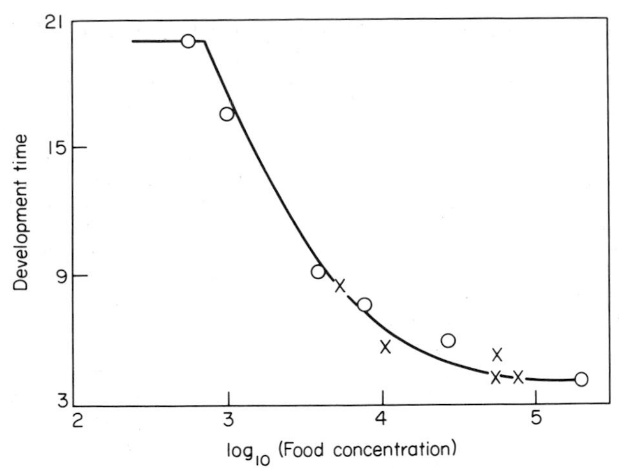

Figure 3. Juvenile development time versus food concentration for *D. pulex* at 20°C. Data from McCauley *et al.*, 1989a (○), Fig. 2 (×).

(C) Assimilation of food, mortality and reproduction

The current 'continuous development' model deliberately neglects the discrete nature of daphnid development, largely on the grounds of technical expediency. Yet our tests of the individual model demonstrated the importance of this aspect, and we require some 'fix' to ensure that we capture the way in which the production of discrete broods smooths out the effect of large, short-term fluctuations in food supply. Similar considerations arise in modelling any mortality due to starvation.

We thus choose to model both reproduction and starvation mortality via an *average* ingestion rate over a time interval equal in duration to one moult, and define

$$\bar{I}_J(t) = T_m^{-1} \int_{t-T_m}^{t} I_J(x)\,dx \qquad \bar{I}_A(t) = T_m^{-1} \int_{t-T_m}^{t} I_A(x)\,dx. \tag{23}$$

To model fecundity, we recognize that neonates released at time t have developed from eggs transferred to the brood pouch at time $t-T_m$. It thus seems appropriate to assume that fecundity depends on $\bar{I}_A(t-T_m)$; indeed we assume that a fixed fraction of all excess assimilate over and above that needed for maintenance goes to reproduction, implying that the instantaneous fecundity is given by

$$\beta(t) = e[\varepsilon_{Aa}\bar{I}_A(t-T_m) - \Gamma_a]. \tag{24}$$

Far less information is available about mortality rates for individual daphnids than is available on growth and reproduction. In particular, we know of no published life tables at different food levels for *D. pulex*, the species we chose to test our model of growth and reproduction. Our representation of mortality thus rests on an interpretation of data for other species (*D. galeata*: Goulden, Henry & Tessier, 1982; *D. magna*: Porter, Orcutt & Gerritsen, 1983), and on preliminary analysis of our own (unpublished) experiments.

We have identified four components of mortality in laboratory populations.
(1) Background.
(2) Senescence.
(3) Starvation.
(4) Production of first brood.

Background mortality is generally agreed to be low. *Senescence* is well documented, there being good evidence (Porter *et al.*, 1983) that life is shorter when food is plentiful. We are currently evaluating evidence that might give pointers to the mechanisms of senescence in *Daphnia*; pending completion of this investigation we have not incorporated senescence explicitly in the model, but instead modify the assumed background mortality rate to give plausible average lifetimes.

To model *starvation mortality*, we again argue that the appropriate determinant is likely to be average food intake over a time comparable with one moult, and on the grounds of parsimony assume dependence on the same averages (equation 23) already used in the model. It is clear from the experiments already cited that starvation mortality is only significant at very low food levels, and this leads us to assume an exponential dependence of the form

$$m_J(t) = m_{oj} + m_{sj} \exp\left[-\bar{I}_J(t)/I_{JO}\right] \tag{25}$$

$$m_A(t) = m_{oa} + m_{sa} \exp\left[-\bar{I}_A(t)/I_{AO}\right] \tag{26}$$

Finally we have included in the model a crude representation of mortality at maturation, based on our own experiments on individual *D. pulex* growing on *Chlamydomonas reinhardii* at densities of 500 cell ml^{-1}. We assumed that any juvenile which has not matured by age T_{max} dies with 100% probability, and that any which mature at age a ($< T_{max}$) have a probability

$$S_{JY}(t) = 1 - \exp\left[-(T_{max} - a)/T_o\right] \tag{27}$$

of surviving the transition to the immature adult stage.

(D) *Qualitative tests of the model*

Laboratory populations

McCauley and Murdoch (1987: see their table 3) reviewed the available literature on laboratory population dynamics and noted that at temperatures around 18–20°C, the expected pattern of behaviour was either damped oscillations (Frank, 1960 for *D. pulex*), or low amplitude cycles (Slobodkin & Richman, 1956 for *D. pulicaria*; Marshall, 1978 and Goulden *et al.*, 1982 for *D. galeata*). The period of the persisting or damped cycles was in the range 20–40 days.

A detailed simulation of any one of these populations requires careful representation of the food replacement schedule, and determination of model parameters appropriate to the species under investigation. However, we might expect to capture the essential features, in particular the period, of the cycles with runs of the continuous development model using our *D. pulex* parameter set and a 'pseudo-chemostat' food replacement schedule in which

$$R_F(t) = D(F_r - F). \tag{28}$$

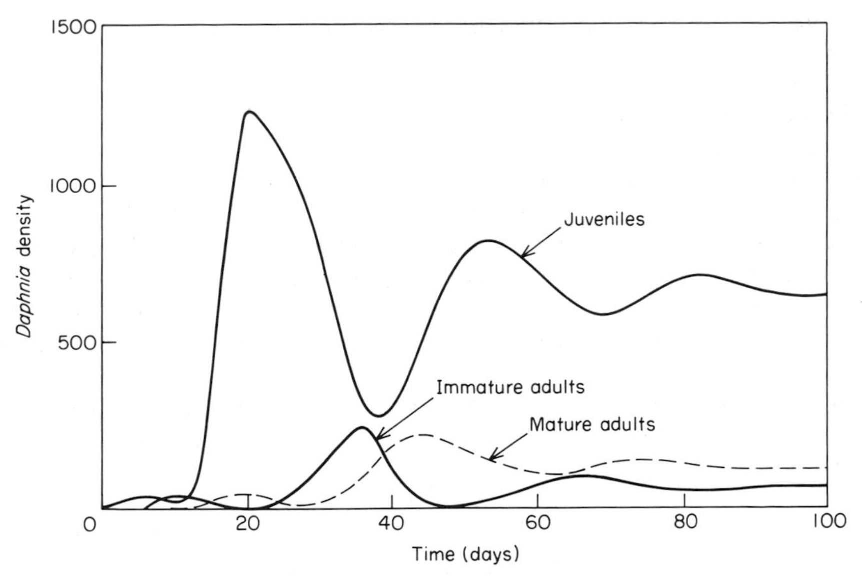

Figure 4. Predicted population dynamics for a 'laboratory' population of *D. pulex*. Details in text.

One such set of results, for a dilution rate of 1 day^{-1} and a reservoir concentration of 1.0 mgC l^{-1} (corresponding to 5×10^4 cells ml^{-1} *Chlamydomonas reinhardii*), is shown in Fig. 4. The cycles are lightly damped, have a period in the desired range, and the demography implied by the stage populations is broadly consistent with observations. However, the detailed form of the cycles is not strictly consistent with the dominance-suppression hypothesis in the Introduction to this paper (A. de Roos, personal communication), an aspect on which we shall report in a future publication.

Field populations

McCauley & Murdoch (1987: see their table 2) also surveyed a number of field populations, and found examples of both 'stable' behaviour and apparent 'prey-predator' cycles, with small amplitudes (ratio of maximum to minimum populations less than four) and essentially the same range of periods as occurred in the laboratory. In examples of prey-predator cycles for which there was information on the size structure of the population, the demography was remarkably similar to that occurring in the laboratory populations.

The difficulties in a realistic simulation of field populations are even greater than those for their laboratory counterparts; the structure of the food assemblage must be considered, and *all* parameters have the potential to vary in response to changing temperature. However, it is again appropriate to investigate the qualitative predictions of our model when we introduce the one key feature that distinguishes the laboratory and field populations: self-reproducing food. We have therefore studied the behaviour of our model with a logistic food replacement function,

$$R_F(t) = rF(1 - F/K). \tag{29}$$

A typical set of results is shown in Fig. 5A and is totally inconsistent with what

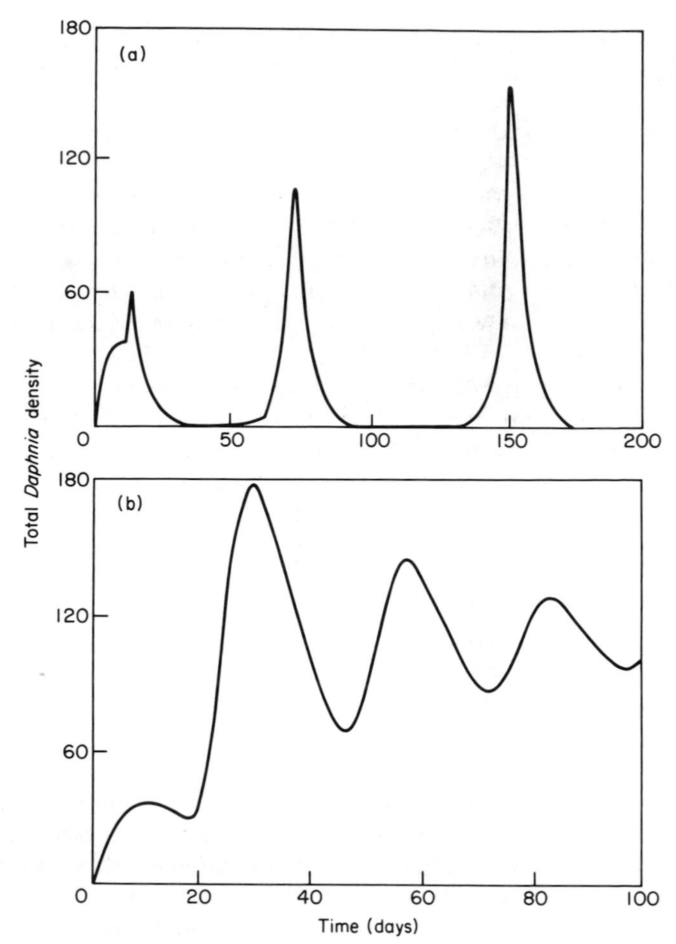

Figure 5. Predicted population dynamics for a *Daphnia* population with reproducing prey as detailed in text: A, using parameters from Table 5, and B, with F_h increased by a factor of 10.

happens in the real world. The ratio of maximum to minimum populations is very large, and the minima are sufficiently small to imply extinction in practice. In addition, the period is far too long. These large amplitude limit cycles are reminiscent of those found in simple, unstructured prey-predator models as a result of the *paradox of enrichment*, in which the combined effect of a saturating functional response for the predator and self-limitation of the prey can be to produce cycles whose very large amplitudes would in practice imply extinction (Gilpin, 1972).

We can look to this literature for guidance on which of our model assumptions and/or parameters is likely to be responsible for the unrealistic cycles. One possibility is that there is some small portion of the food inaccessible to the *Daphnia* at any given time. Such a refuge might arise if *Daphnia* avoids the top 1–2 m of lake water because of high light levels and hence high visibility to potential predators (Gulati, 1978), or on a horizontal scale in stratified systems because of Langmuir circulations (George & Edwards, 1973). We therefore performed some simulations with *constant number refuges*. Even a refuge as low as

5% of carrying capacity is sufficient to dampen out the large amplitude prey-predator cycles, interestingly leaving behind 'single-generation cycles' with a period of the same order as that observed.

A second, and perhaps more plausible, mechanism is suggested by work of McCauley *et al.* (1988) on factors determining the equilibria of natural *Daphnia* populations. There, it was argued that the model parameters were likely to be strongly influenced by the structure of the phytoplankton assemblage, and in particular by variations in the edible fraction of the algal population. It was further argued (by reference to the daphnid's mode of eating) that the introduction of inedible food would not influence the handling time per unit of edible food, but would reduce the effective filtering rate. In terms of the parameters in Table 4, this would imply that the half-saturation constant F_h is too low. We therefore performed a set of runs in which the value of this parameter was varied, one result being reproduced as Fig. 5B. Again we see the disappearance of the large-amplitude, long-period cycles and the occurrence of low-amplitude, single-generation cycles.

DISCUSSION

The work outlined in this paper has drawn attention to the large number of biological judgements that are necessary when constructing an individual model, even for a genus as well studied as *Daphnia*. We arrived at an individual model that is inelegant and parameter-rich, but is consistent with a rather wide body of experimental data. Clearly an important component of future work should be a search for simpler individual models which retain the capacity to predict growth and fecundity over a wide range of food densities and in rapidly varying conditions.

Further study is also needed on some technical problems associated with the transition from individual to population models. Our 'continuous development' population model does appear to be a useful simplification, but we would feel safer using it if we knew how its properties compare with those of a model rigorously derived from our individual model.

However, the central, still unanswered question is this: can structured population models help us understand the factors responsible for determining population sizes and patterns of fluctuations in natural populations? The approach to population modelling adopted in this paper is deliberately unbalanced: we work from a detailed account of the individual behaviour of members of the population under investigation, yet include as little detail as possible on the environmental factors (such as 'food') that drive the population dynamics. We do this, fully recognizing that real, natural populations are spatially heterogeneous, and form part of a multi-species community in which the dynamics of each species is potentially influenced by many others. The broad justification is that this approach opens the possibility of quantifying the relative importance of the many factors that *may* influence the population dynamics. A lake may not be merely a 'scaled-up' litre flask, but the analyses of McCauley & Murdoch (1987) suggest that *Daphnia* populations may behave similarly in both systems; structured models can sharpen our understanding of the ways in which they differ, thereby contributing significantly to our understanding of natural populations.

ACKNOWLEDGEMENTS

This work has been supported by grants from S.E.R.C., N.S.E.R.C., N.A.T.O., and the British Council. We thank Andre de Roos for critical comments on the first draft of this paper.

REFERENCES

FRANK, P. W., 1960. Prediction of population growth form in *Daphnia pulex* cultures. *American Naturalist, 94:* 357–372.

GEORGE, D. G. & EDWARDS, R. W., 1973. *Daphnia* distribution within Langmuir circulations. *Limnology & Oceanography, 18:* 798–800.

GILPIN, M. E., 1972. Enriched predator-prey systems: theoretical stability. *Science, NY, 177:* 902–904.

GOULDEN, C. E., HENRY, L. L. & TESSIER, A. J., 1982. Body size, energy reserves and competitive ability in three species of Cladocera. *Ecology, 63:* 1780–1789.

GULATI, R. D., 1978. Vertical changes in the filtering, feeding, and assimilation rates of dominant zooplankters in a stratified lake. *Verhandlungen, Internationale Vereinigung für Theoretische und Angewandte Limnologie, 20:* 950–956.

GURNEY, W. S. C., NISBET, R. M. & BLYTHE, S. P., 1986. The systematic formulation of models of stage-structured populations. In J. A. J. Metz & O. Diekmann (Eds), *The Dynamics of Physiologically Structured Populations*. Heidelberg: Springer Verlag.

GURNEY, W. S. C., NISBET, R. M. & LAWTON, J. H., 1983. The systematic formulation of tractable single-species models incorporating age structure. *Journal of Animal Ecology, 52:* 479–495.

GURNEY, W. S. C., NISBET, R. M., McCAULEY, E. & MURDOCH, W. W., 1989. The physiological ecology of *Daphnia*. II. Formulation and tests of a dynamic model of growth and reproduction. *Ecology*, in press.

INGLE, L., WOOD, T. R. & BANTA, A. M., 1937. A study of the longevity, growth, reproduction and heart rate in *Daphnia longispina* as influenced by limitations in the quantity of food. *Journal of Experimental Zoology, 76:* 325–352.

KOOIJMAN, S. A. L. M., 1986a. Population dynamics on the basis of budgets. In J. A. J. Metz and O. Diekmann (Eds), *The Dynamics of Physiologically Structured Populations*. Heidelberg: Springer Verlag.

KOOIJMAN, S. A. L. M., 1986b. Energy budgets can explain body size relations. *Journal of Theoretical Biology, 121:* 269–282.

LAMPERT, W., 1975. A tracer study on the carbon turnover of *Daphnia pulex*. *Verhandlungen, Internationale Vereinigung für Theoretische und Angewandte Limnologie, 19:* 2913–2921.

LYNCH, M., WEIDER, L. J. & LAMPERT, W., 1986. Measurement of the carbon balance in *Daphnia*. *Limnology & Oceanography, 31:* 17–33.

MARSHALL, J. S., 1978. Population dynamics of *Daphnia galeata mendotae* as modified by chronic cadmium stress. *Journal of the Fisheries Research Board of Canada, 35:* 461–469.

McCAULEY, E. & MURDOCH, W. W., 1987. Cyclic and stable populations—plankton as paradigm. *American Naturalist, 129:* 97–121.

McCAULEY, E., MURDOCH, W. W. & WATSON, S., 1988. Simple models and variation in plankton densities among lakes. *American Naturalist, 132:* 383–403.

McCAULEY, E., MURDOCH, W. W. & NISBET, R. M., 1989a. Growth, reproduction and mortality of *Daphnia pulex*: life at low food. Submitted to *Functional Ecology*.

McCAULEY, E., MURDOCH, W. W., NISBET, R. M. & GURNEY, W. S. C., 1989b. The physiological ecology of *Daphnia*. I. Development of a model of growth and reproduction. *Ecology*, in press.

METZ, J. A. J. & DIEKMANN, O., (Eds), 1986. *The Dynamics of Physiologically Structured Populations*. Heidelberg: Springer Verlag.

MURDOCH, W. W. & McCAULEY, E., 1985. Three distinct types of dynamics shown by a single planktonic system. *Nature, London, 316:* 628–630.

NISBET, R. M. & GURNEY, W. S. C., 1983. The systematic formulation of population models for insects with dynamically varying instar duration. *Theoretical Population Biology 23:* 114–135.

PALOHEIMO, J. E., CRABTREE, S. J. & TAYLOR, W. D., 1982. Growth model of *Daphnia*. *Canadian Journal of Fisheries and Aquatic Science, 39:* 598–606.

PETERS, R., 1983. *The Ecological Implications of Body Size*. Cambridge: Cambridge University Press.

PORTER, K. G., ORCUTT, JR, J. D. & GERRITSEN, J., 1983. Functional response and fitness in a generalist filter feeder, *Daphnia magna* (Cladocera: Crustacea). *Ecology, 64:* 735–742.

RICHMAN, S., 1958. The transformation of energy by *Daphnia pulex*. *Ecological Monographs, 28:* 273–291.

SLOBODKIN, L. B. & RICHMAN, S., 1965. The effect of removal of fixed percentages of the newborn on size and variability of *Daphnia pulicaria* (Forbes). *Limnology & Oceanography, 1:* 209–237.

TAYLOR, B. E., 1985. Effects of food limitation on growth and reproduction of *Daphnia*. *Archiv fuer Hydrobiologie Beihatrage, 21:* 285–296.

TAYLOR, B. E. & GABRIEL, W., 1985. Reproductive strategies of two similar *Daphnia* species. *Verhandlungen, Internationale Vereinigung für Theoretische und Angewandte Limnologie, 22:* 3047–3050.

Biological Journal of the Linnean Society (1989), *37:* 101–116. With 8 figures

A life-cycle theory of responses to stress

R. M. SIBLY

Department of Pure and Applied Zoology, University of Reading, Reading RG6 2AJ

AND

P. CALOW

Department of Animal and Plant Sciences, University of Sheffield, Sheffield S10 2TN

Stress is here defined as an environmental condition that reduces Darwinian fitness when first applied. Optimal stress responses (i.e. those that maximize Darwinian fitness) are calculated for different levels of growth and mortality stress, and are found to depend critically on the shape of the trade-off curve relating mortality to growth rate. If the trade-off does not change shape when stress is applied, then the optimal strategy is to spend less on personal defence for both mortality and growth stresses. However, if stress does change the shape of the trade-off the predictions may be modified, or reversed. This optimality analysis is rigorous and easy to apply. What is more difficult, is to establish the shapes and positions of trade-off curves in particular cases. This problem is discussed and some suggestions are made. The theory's predictions are applied speculatively to biogeographical data on marine animals and are found to be qualitatively successful, although some of the needed data are lacking. The applications and testability of the theory in the study of ageing and a variety of other processes are considered.

KEY WORDS:—Stress – growth – mortality – optimality theory – Darwinian fitness.

CONTENTS

INTRODUCTION

In this paper we use the optimality approach to examine how animals are likely to respond to different levels of stress. The basis of the method is to locate

0024–4066/89/050101+16 $03.00/0

the strategy (i.e. life cycle) that maximizes Darwinian fitness, subject to various constraints that arise because an animal's options are limited. Identification of these options, and the trade-offs that arise from them, is crucial to understanding the likely outcomes of the evolutionary process (Sibly & Calow, 1986).

We define stress very generally as an environmental condition that, when first applied, impairs Darwinian fitness; for example, reduces survivorship (S) and/or fecundity (n) and/or increases the time (t) between life-cycle events. Some other definitions have been more restrictive, but are still a subset of this. For example, Grime (1979) uses the term 'stress' only to describe factors that inhibit production (i.e. influence n and t) and uses the term 'disturbance' to describe factors that impair survivorship (i.e. S). Instead we refer to these respectively as growth and mortality stress, and these are defined more precisely below.

The effect of a mortality stress may be alleviated to some extent if the organism spends more of its metabolic income on defence. However, what is spent on defence is not available for growth, so there may be a trade-off between mortality and growth. By the same token, the effects of a growth stress may be alleviated by spending more on growth and less on defence.

The trade-off between mortality and growth is a central concept in this paper and limits the organism's options. We shall consider which option represents the optimal strategy, and then ask how it is affected by mortality and growth stresses. The presumption is that stress acts as a selection pressure and that, given appropriate circumstances, organisms can respond adaptively to it (e.g. as in the response of plants to pollution stress; Bradshaw & McNeilly, 1981). In later sections we evaluate the predictions by reference to marine biogeographical data. We give examples to show how the trade-off curve may be measured in practice, and we consider the application of the theory to the study of ageing and other processes.

THE TRADE-OFF CURVE BETWEEN MORTALITY AND GROWTH

This could occur for behavioural reasons—a food-limited animal can sometimes increase its growth rate by foraging in more dangerous places, or at more dangerous times of day, so increasing growth rate at the cost of increased mortality rate. Or it could occur through investment in defensive structures such as spines: for example, sticklebacks invest in spines to defend themselves against predators, but presumably at the expense of growing faster. Another type of defence consists in laying down reserves against times of nutrient or water shortage, as in desert plants. Similarly some fledgling birds invest in fat reserves, at the cost of a slower growth rate (Lack, 1968; O'Connor, 1984; Sibly & Calow, 1987a).

Many examples of personal defence, such as escaping predators, neutralizing and excreting toxic chemicals, or pumping out excess water or ions, consist of active processes that consume power and hence energy. Protein synthesis and turnover are metabolically costly, accounting for 16% basal metabolic rate in the marine bivalve *Mytilus edulis* (Hawkins, 1985) and more than 15% total energy expenditure in mammals (Garlick, Burk & Swick, 1976; Nicholas, Lobley & Harris, 1977; Garlick, 1980; Waterlow, 1980; Meier *et al.*, 1981). Hence, the replacement and repair of proteins damaged by intrinsic processes and/or toxic chemicals and high-energy radiation are likely to be expensive (but cf. Koehn &

Bayne, 1989). Similarly insects that cope with toxins may continuously produce a variety of enzymes to degrade foreign compounds (Oppenoorth, 1985) or synthesize additional enzymes only when required (Terrierc, 1984). Likewise, combatting disease organisms by deploying an immune system involves, amongst other things, extensive protein synthesis and so is also likely to be metabolically costly.

There are, therefore, good *a priori* grounds for believing that defence is generally expensive in power and energy. Given a finite supply of energy and resources available for metabolism, it follows that if more energy is invested in defence less will be available for production. Thus 'scope for growth' (i.e. energy available for production) is lower in marine mussels (*Mytilus edulis*) at sites exposed to heavy metals than those at 'cleaner' sites (Bayne *et al.*, 1979) and is reduced in animals having greater tissue loads of oil-derived hydrocarbons (Moore *et al.*, 1988). This variation appears to be largely an environmental, phenotypic effect, but there is some suggestion of a genetic component (Widdows *et al.*, 1984). Similarly, heavy-metal tolerant ecotypes of several species of flowering plants have lower production rates than non-tolerant ecotypes when grown in normal soils and this difference is greatest in mixed as compared with pure stands (Cook, Lefebvre & McNeilly, 1972; Wilson, 1988). In the same way, the trade-off observed between individual growth rate and survival rate for a variety of species of fishes (Beverton & Holt, 1959) and sea urchins (Ebert, 1982, 1985) could arise because some species invest more in 'self preservation' at the expense of growth.

The trade-off can, therefore, be genetically based, and this is a likely outcome if one strain, say A, is fitter in environment A, and another strain, B, is fitter in environment B. Field data comparing resistant and susceptible strains of various animals in stressed and unstressed environments (Table 1) indicate that resistant strains are fitter in stressed environments, and susceptible strains fitter in unstressed environments. Note, however, that whereas resistants are often much fitter than susceptibles in the stressed environment, the fitness of the resistants in the unstressed environmnent is sometimes not much less than that of susceptibles (e.g. 0.95 *vs.* 1).

MODELS AND DEFINITIONS

Fitness (F) of a dominant gene in a specified environment is measured as its per capita rate of increase, and this is given by:

$$1 = \tfrac{1}{2}ne^{-(\mu_j+F)t_j} + e^{-(\mu_a+F)t_a} \quad (1)$$

where males and females carrying the gene have life cycles with the following parameters: t_j = time between birth and first breeding; t_a = time between breeding seasons; μ_j = mortality rate for juveniles; μ_a = mortality rate for adults; n = fecundity each breeding season (Sibly & Calow, 1986; Sibly, 1989). (Strictly the expression applies only to absorption-costing organisms but the results of the analysis that follows apply also to direct-costing organisms, *sensu* Sibly & Calow, 1984.) Although we develop the analysis specifically for the trade-off for juveniles between μ_j and t_j, an exactly analogous analysis applies to the trade-off for adults between μ_a and t_a.

TABLE 1. *Fitness of resistant strains in environments with/without pesticides. Fitness is here the population genetics measure, and is relative to the fitness of susceptibles, which is taken to be 1*

Species	Pesticide	Fitness with pesticide	Fitness without pesticide
1 Mosquito, *Anopheles culifacies*	DDT	1.3–1.5	0.62–0.97
Mosquito, *Anopheles culifacies*	Dieldrin	2.9–6.1	0.44–0.79
2 Rat, *Rattus norvegicus*	Warfarin	$\gg 1$	0.54
3 Moth, *Plodia interpunctella*	Virus	> 10	0.95

Data from: 1, Wood & Bishop (1981). See Roush & McKenzie (1987) for a review of insecticide resistance; 2, Bishop, 1981; 3, Vigneswaren, Hunter-Fujita & Sibly (unpublished).

In considering the effects of stresses it is convenient to define growth rate (g) as follows:

$$g = k/t_j, \tag{2}$$

where k is a constant.

Therefore:

$$1 = \tfrac{1}{2}ne^{-(\mu_j+f)K/g} + e^{-(\mu_a+F)t_a}. \tag{3}$$

Rearranging:

$$\mu_j = g\left[\frac{1}{k}\log\frac{n}{2} - \frac{1}{k}\log\left(1-e^{-(\mu_a+F)t_a}\right)\right] - F. \tag{4}$$

For given values of n, μ_a, F and t_a the term in square brackets is constant so:

$$\mu_j = g \times \text{constant} - F. \tag{5}$$

Thus in a plot of μ_j versus g (Fig. 1A), lines of equal fitness are straight, and increasing fitness is represented by the lines lower down the μ_j axis. Note also that the line for zero fitness must pass through the origin (see below, p. 107).

We suppose that throughout the juvenile period there is a trade-off between growth and mortality (as above). Furthermore, we presume that the curve is convex seen from below as in Fig. 1B. The rationale for this is that (a) though the ability to actively pump toxic materials (e.g. ions and water) and the synthesis of new proteins is likely to be linearly related to ATP availability (cf. Potts & Parry, 1964), (b) the effect of increasing concentration of stress or the damage it causes is likely to become increasingly serious (see example in Fig. 7, below).

Since ATP availability for defence is likely to be proportional to the difference between energy input (I) and the energy used in growth (G), it follows that μ is likely to increase increasingly as $I-G$ reduces and hence as G increases.

We are now in a position to calculate the optimal (maximum F) strategy, by superimposing lines of equal fitness (Fig. 1A) onto the trade-off curve (Fig. 1B), as in Fig. 1C. The optimal strategy in an unchanging population ($F = 0$; see below) is the point where the trade-off curve touches the straight line through the origin. Although the method of calculating the optimal strategy (i.e. life cycle) has been worked out here for juveniles, exactly the same method can be applied to adults collecting resources for breeding.

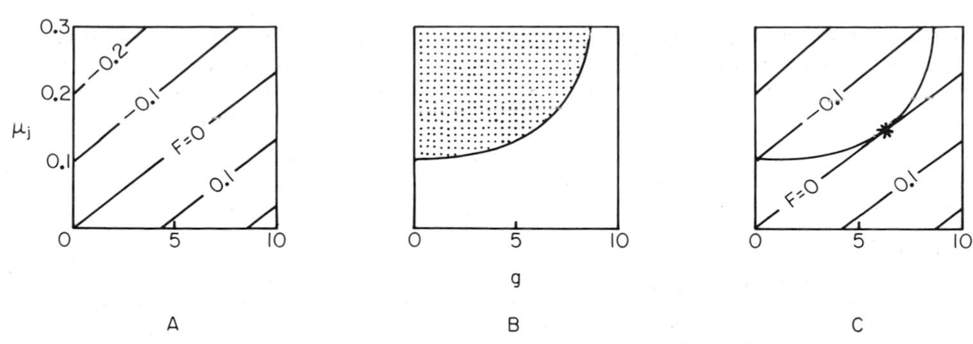

Figure 1. A, Lines of equal fitness, F, in the $g\mu_j$ plane: g = growth rate, mg d^{-1}; μ_j = mortality rate, d^{-1}. Parameter values were k = 100 mg, $\mu_a = 0.3\ d^{-1}$, n = 20, $t_a = 10$ d. B, The animal's options are represented by the stippled set. The boundary of this set is the trade-off curve relating mortality rate to growth rate. The curve illustrated is convex seen from below. C, Superimposing A and B allows identification of the optimal strategy, *. See text for further details.

Following the definitions given above, we define a stress as any environmental change that worsens the position of the trade-off curve by shifting it towards a lower fitness value and we distinguish different types of stress on the basis of the direction in which they move the trade-off curve. Thus a mortality stress simply increases the mortality rate by shifting the trade-off curve vertically upwards (Fig. 2). This may or may not change its shape—see below. A growth stress shifts the trade-off curve horizontally to the left, again, with or without a shape change. In Grime's (1979) terms our mortality stress is equivalent to 'disturbance' and growth stress to 'stress'. An example of a 'pure' mortality stress might be a predator, and of a 'pure' growth stress a reduction, within the physiological range, of temperature for poikilotherms and possibly PO_2 for aquatic animals. Other stresses, such as many chemical toxins, might simultaneously reduce the survival chances and the growth rates of organisms exposed to them.

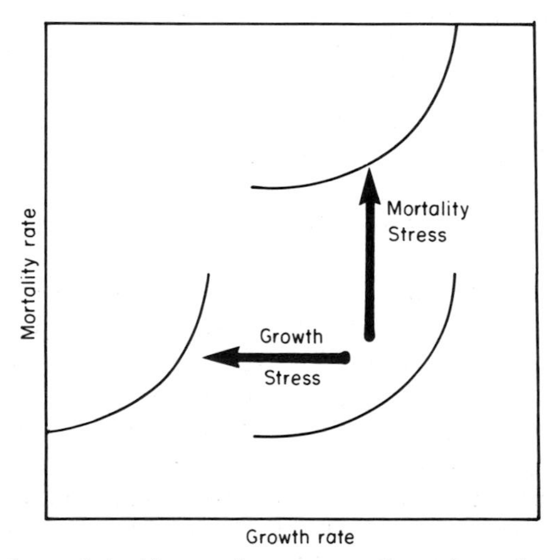

Figure 2. The distinction made in this paper between mortality and growth stresses. Mortality stress moves the trade-off curve vertically upwards and growth stress moves it horizontally to the left.

However, to gain initial insight into the effects of stress we shall consider first simple vertical and horizontal displacements of the trade-off curve within the $g\mu$ plane.

STRESSES THAT DO NOT CHANGE THE SHAPE OF THE TRADE-OFF CURVE

Suppose that the resources available to the animal are split between growth g and defence d, so that (available resources) = g+d. All these quantities are rates. Let μ_0 represent basal mortality rate, i.e. the effect of stresses that cannot be counteracted however much is spent on defence, and let $\mu_1(d)$ represent the additional 'defendable' mortality rate that results when resources are spent on defence at rate d. Total mortality rate $\mu(d)$ is then given as $\mu(d) = \mu_0 + \mu_1(d)$. Alternatively μ may be written as a function of g, $\mu(g)$. These ideas can be represented diagrammatically (Fig. 3). It can be seen that μ_0 represents the asymptotic value of the $\mu(d)$ curve as $d \rightarrow \infty$.

If the basal and defendable mortality agents are statistically independent, or act in sequence, then increasing μ_0 does not affect the value of μ_1 (d). This is because survivorship over time period Dt, which we shall call S(Dt), is then the product of survivorship of basal and defendable mortality, $S_0(Dt)$ and $S_1(Dt)$, where

$$S_0(Dt) = e^{-\mu_0 Dt} \quad \text{and} \quad S_1 = e^{-\mu_1 Dt}$$

and therefore

$$S(Dt) = S_0(Dt).S_1(Dt) = e^{-\mu_0 Dt - \mu_1 Dt}.$$

But S(Dt) is equal to $e^{-\mu(d)Dt}$, and therefore $\mu(d) = \mu_0 + \mu_1(d)$ still holds even if the value of μ_0 is changed. So if the mortality agents are statistically independent or act in sequence, the effect of mortality stress is to shift the $\mu(g)$ curve vertically upwards, while maintaining its shape. The amount by which the curve is shifted upwards represents the additional mortality rate introduced by the mortality stress.

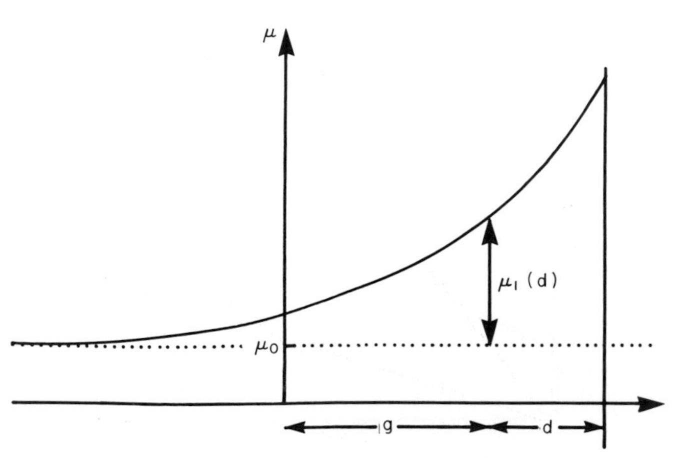

Figure 3. The relationship between expenditure on defence, d, the 'defendable' mortality rate, $\mu_1(d)$; and the $\mu(g)$ curve. See text for further details.

Applying a growth stress does not affect μ_0, and we will assume it does not affect the dependence of μ_1 on defence expenditure, but it does reduce the available resources, say by Q. Growth rate is then reduced by the same amount (i.e. Q) if defence expenditure is held constant. In other words a growth stress moves the $\mu(g)$ curve horizontally to the left, while maintaining its shape.

During juvenile growth the situation is complicated by the increase in resource availability as the animal grows, and the shape of the trade-off curve may also change. However, the above analysis is still likely to apply over a short time period. We also have to take account of the Principle of Ecological Compensation, according to which mortality, growth and birth rates cannot vary independently between stable populations (Sibly & Calow, 1986, 1987b). Thus most populations are at steady state ($F = 0$) and have to return to this state after stress, otherwise they become extinct. It follows that if either juvenile growth or mortality rate is stressed, and F remains at zero, there have to be adjustments in n and/or μ_a and/or t_a, possibly by density-dependent ecological factors. For example, a higher μ_j means a lower density of juveniles, and this in turn might mean more food for adults and hence a higher reproductive output. To meet the requirement that $F = 0$, the fitness contours have to pass through the origin (this follows from equation (5)). So the optimal strategies in all cases occur where the trade-off curves touch straight lines through the origin, and are found by inspection (Fig. 4).

Predictions

The cases illustrated in Fig. 4 indicate that if growth or mortality stresses operate in the way defined in the last section, i.e. influencing the position but not the shape of the trade-off curve, then both favour less investment in defence; i.e. optimal growth rates and mortality rates should increase. Hence populations exposed to poor environmental conditions for growth should invest less in defence and repair (see below) and this will also be the case for populations

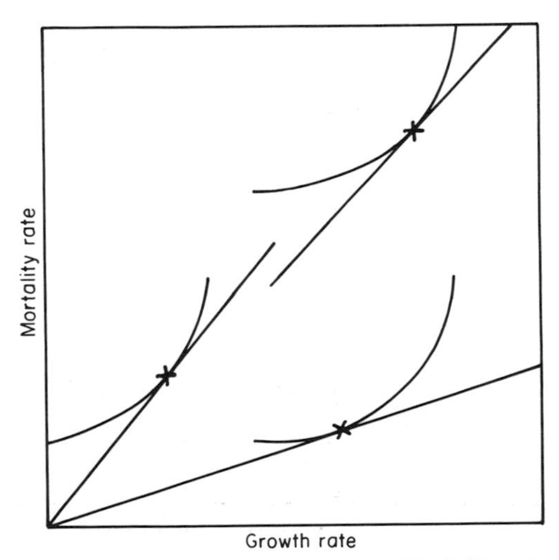

Figure 4. Optimal strategies for the three curves illustrated in Fig. 2. Note that most should be spent on defence (i.e. least on growth) in the absence of mortality and growth stresses.

exposed to high levels of mortality. Under conditions of resource restriction, less should be invested in defence. Similarly, if high mortality rates apply, a short generation time is favoured since this will reduce the probability of death between birth and breeding.

Whereas the optimality analysis (above) is rigorous and relatively easy to carry out, the arguments in the rest of the paper are more speculative. This is because there is a lack of good data on the shapes and positions of trade-off curves in particular cases.

Implications for defences against predators and community structure

It follows from the previous section that as conditions for growth (production) improve then a bigger investment in defence is favoured, provided that the shape of the trade-off curve is unchanged. The clearest clines of growth stress probably occur in the seas, and we therefore turn to marine animals to examine the hypothesis that less is spent on defence where growth is slowest. We shall assume in what follows that there are clines of growth stress with latitude and with height up the shore, and furthermore that growth rates are generally slower towards the poles, where it is colder; and higher in the intertidal zone, because feeding periods are shorter. The first assumption depends on the availability of food being constant, or not varying too much, along the cline. Should data become available which falsify either of these assumptions, then the following arguments will need revision.

The investment in defence along each cline can be examined using the biogeographical data collated by Vermeij (1978). He considers, for example, that gastropod shells have greater mechanical resistance to crushing both towards the equator, and as one goes down the shore on most rocky coasts. This is achieved in part by coarser shell structure, with relatively larger knobs and spines. Similarly, bivalves nearer the equator have sturdier valves which completely enclose the soft parts, giving greater protection against predators which crush, drill or forcibly open valves. Vermeij concludes that, in general, antipredatory defences are inversely correlated with what we here call growth stress, along both of these clines, in all taxonomic groups for which data are available. These include gastropods, bivalves, barnacles, sponges, holothurians and marine sessile photosynthesizers.

One possible complication discussed by Vermeij arises because calcification is probably more costly at lower temperatures, and this will also be true of freshwater systems. The effect of this on our model would be to make the trade-off curve shallower at lower temperatures or in freshwater environments, and the optimal strategy would therefore 'move to the right', even further than it would have done under the influence of growth stress alone. Thus predictions should be exaggerated by this complication. Sculpture is certainly markedly reduced in freshwater gastropods and bivalves (see respectively Vermeij & Covich, 1978; Vermeij & Dudley, 1975) as compared with marine relatives.

We suggest, therefore, that as organisms become more defended, predators have to adapt by coevolution and hence become more specialized. Hence elaborateness of defence structures and the diversity of predators should improve with environmental productivity—as they do.

However, there are alternative hypotheses. The armament and probably the effectiveness of predators covary with prey defences, which we interpret as an evolutionary adaptation on the part of predators. By contrast Vermeij (1978) suggests that these predator characteristics are the cause of the biogeographical associations. To explain the variations in predator characteristics, he suggests that predators in warmer waters possess increased muscular capability, leading to an increase in predator armament. In addition, predator effectiveness may be reduced away from the equator because they have to survive seasonal troughs in food abundance. Temporal restrictions could also limit the effectiveness of upper shore predators. Vermeij supposes that where predators are more effective, prey respond with increased defence.

Another possible explanation of the biogeographical data is that increased productivity will increase the biomass of prey and this makes increased niche separation possible in predators. Increased diversity of predators leads by coevolution to increased diversity of prey (a variant of the Productivity Hypothesis of Connell & Orians, 1964). The difference between these hypotheses is illustrated schematically in Fig. 5. All three predict increasing diversity/elaborateness of defences and predator diversity with environmental productivity. However the Productivity Hypothesis and Vermeij's Hypothesis assume that predator armament came first, and that prey defence was increased as an evolutionary response, whereas the Defence-Release Hypothesis assumes the reverse. Hence, because of the sequence of causality one might expect the Productivity Hypothesis and Vermeij's Hypothesis to yield better correlations between diversity of predators and environmental productivity than between diversity of prey/defence structures and environmental productivity, and the reverse for the Defence-Release Hypothesis. This therefore points to a possible way of distinguishing between the hypotheses.

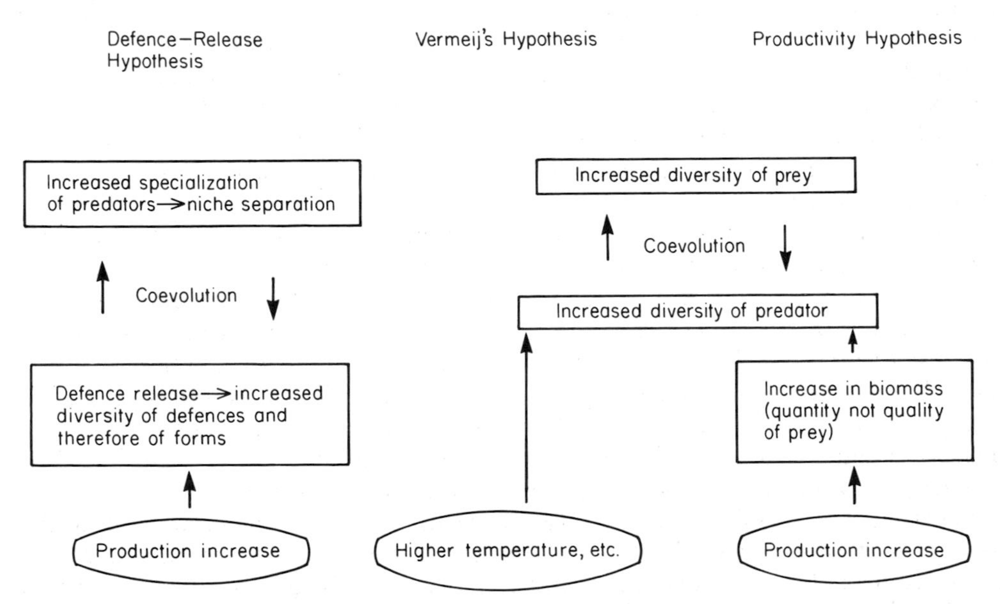

Figure 5. The rival hypotheses for the explanation of Vermeij's biogeographical data.

STRESSES THAT CHANGE THE SHAPE OF THE TRADE-OFF CURVE

Stresses that change the shape of the trade-off curve $\mu(g)$ can select for increased not reduced investment in defence. Two cases are pertinent and are illustrated in Fig. 6. In A the trade-off disappears in the absence of mortality stress. An example would be the complete disappearance of predators, or toxins. As shown on the left of the figure, slow growers are well-defended, so their growth and mortality rates are the same with or without a predator. On the other hand fast growers are taking a big risk if there are predators around. In the absence of predators there is nothing lost by growing fast, dispensing with all defences, because animals growing fast obtain mortality rates just as low as those that grew slowly. As a result the options set becomes rectangular, and the optimal strategy (starred) is to grow as fast as possible. In this case the earlier predictions are reversed, i.e. less (not more) should be spent on defence at low levels of mortality stress—because defence brings no corresponding benefits in terms of reduced mortality. This provides a likely explanation for the evolution of flightless birds with vestigial wings on islands, and for the evolution of eyeless, uncoloured animals in caves.

It is also clear that for populations exposed to pollutants there could be evolution of mechanisms to resist their toxic effects (Bradshaw & McNeilly, 1981), but these are likely to be energy expensive (see above) and hence take place only at the expense of somatic growth (Cook *et al.*, 1972). This therefore represents the converse case to that described above in Fig. 6A. Pollution causes the trade-off curve to move upwards in the $g\mu_j$ plane but more so for fast than slow growers. Here the optimum solution is to evolve resistance and to grow more slowly. The extent to which this kind of response can evolve will depend upon the extent to which there is genetic variance for the resistance trait and the extent to which it confers advantages in μ_j relative to reductions in g (i.e. how much it tilts the trade-off curve).

In Fig. 6B the trade-off is alleviated in the absence of growth stress. An example would be the provision of a previously limiting resource. There is then little lost by reducing mortality as much as possible, since growth rate is not

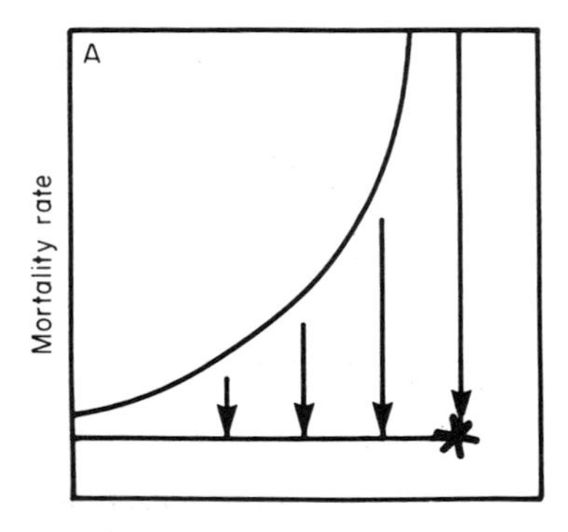

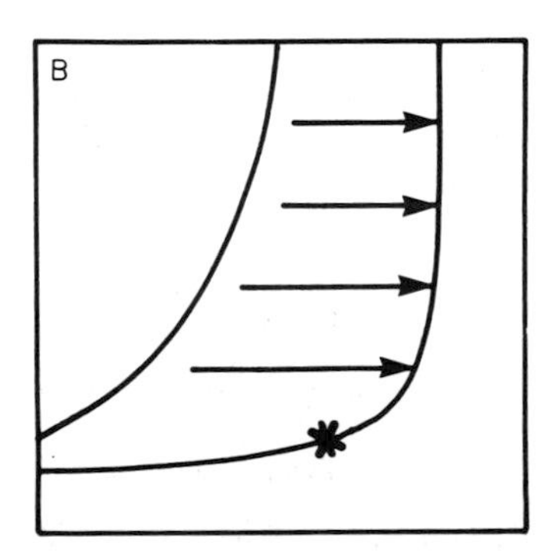

Growth rate

Figure 6. Stresses that change the shape of the trade-off. Arrows show the effects of removing A, mortality and B, growth stress. In the absence of mortality stress, A, the optimal strategy moves to the right (starred)—reversing the prediction of Fig. 4. In the absence of growth stress, B, the optimal strategy again moves to the right, implying lower expenditure on defence. Again, this is the reverse of Fig. 4.

thereby much reduced. Again the trade-off curve becomes nearly rectangular, and the optimal strategy is to reduce mortality rate to the lowest possible level, but this is now compatible with fast growth and low expenditure on defence. In this case also earlier predictions are reversed.

MEASUREMENT OF THE $\mu(g)$ CURVE

Most of the data needed to construct the $\mu(g)$ curve are available in a study of detoxification of metals by the springtail *Onychiurus armatus* (Collembola) (Bengtsson, Gunnarsson & Rundgren, 1983, 1985). The springtails were fed on fungi grown on a nutrient broth contaminated with 0, 30, 90 or 300 $\mu g\ g^{-1}$ of copper and lead in equal proportions. Concentrations of these metals within the animal reached high levels initially, but were then reduced by detoxification processes reaching steady state after a few weeks (Fig. 7A). Detoxification was achieved by moulting more frequently, and this reduced the growth rate, as shown in Fig. 7B. Detoxification was not complete, however, so that body metal

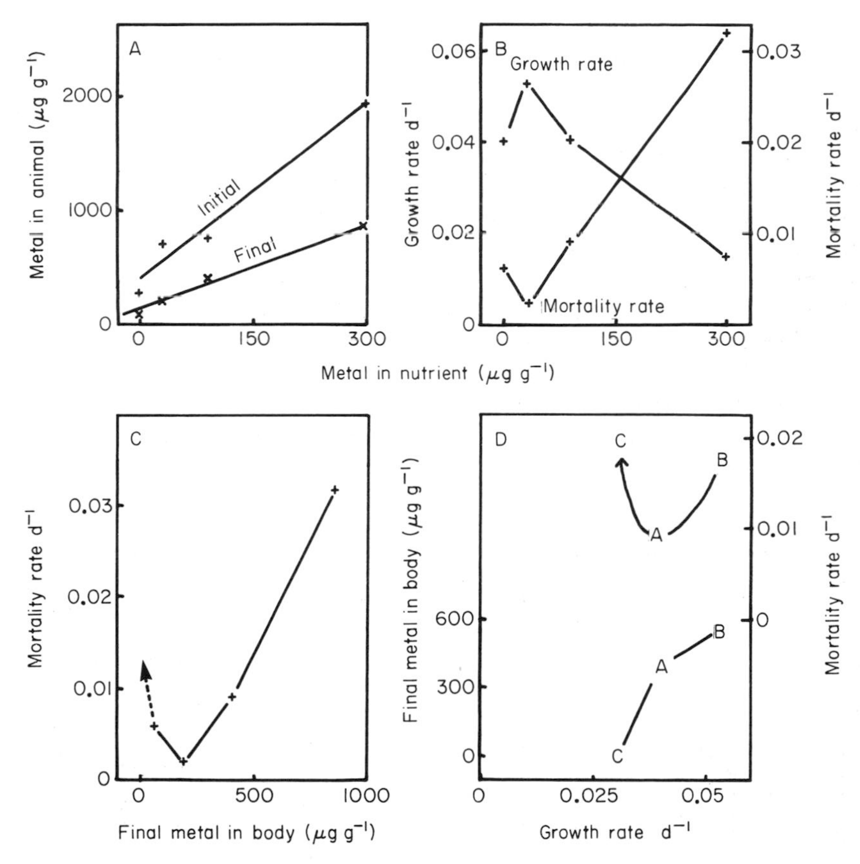

Figure 7. A, Initial peak levels and final steady-state levels of metals (Cu and Pb) in the body in relation to levels in the nutrient broth. B, Growth rate (reciprocal of time to first reproduction which occurs at body length 0.9 mm) and mortality rate (instantaneous daily rate calculated from survivorship over the first 10 weeks of life). C, Mortality rate in relation to the final steady-state levels of metals in the body. Mortality must increase at very low levels (dashed line) as the animals then suffer from copper deficiency. D, The options as regards growth rate and mortality rate when metal levels in the nutrient broth are 90 $\mu g\ g^{-1}$. All data are from graphs in Bengtsson, Gunnarson & Rundgren (1983, 1985).

levels were elevated in more contaminated environments even at steady state, as shown in Fig. 7A. Mortality rate was higher in the contaminated environments (Fig. 7B), and we shall assume that this was the result of the increased metal levels in the body, as shown in Fig. 7C. Note, however, that a certain amount of copper is needed physiologically, so that in a completely deficient environment growth rate must be reduced and mortality rate increased (Fig. 7B).

We are now in a position to estimate the shape of the $\mu(g)$ curve. Consider the 90 $\mu g\ g^{-1}$ environment. We know that the actual growth rate achieved was 0.04 d^{-1}, and that this resulted in a final metal level in the body of 400 $\mu g\ g^{-1}$, and hence a mortality rate of 0.009 d^{-1} (points A in Fig. 7D). What other options are there? If growth rate had been a maximum (Points B) then less detoxification would have been possible—how much can be estimated from Fig. 7A, assuming the illustrated reductions in metal levels are entirely due to growth rate.

This allows us to calculate the final metal level in the body and hence the mortality rate (Points B in Fig. 7D). On the other hand the same method suggests that if growth rate had been a minimum the detoxification processes would have removed the entire metal contents of the body, leading to increased mortality (Point C in Fig. 7D). The upper curve in Fig. 7D is the required $\mu(g)$ relationship. As expected the curve is convex viewed from below. It does not asymptote to the horizontal at low growth rates, as in Fig. 3, but this is because at low levels, metals are necessary for survival.

The $\mu(g)$ curve can be constructed more easily for the aphid *Myzus persicae* detoxifying insecticides, because strains are available that differ in their insecticide resistance (Devonshire & Sawicki, 1979; Oppenoorth, 1985). The different options are genetically coded, and the mortality and growth consequences of each option can be examined in a standard environment. More resistant strains contain more detoxifying enzyme (phosphatase E_4), possibly because they contain more duplications of a structural gene (Fig. 8A). LD_{50}s have been measured for three of these strains, from which we have backcalculated mortality rate (Fig. 8B). As expected the curve relating mortality rate to defence expenditure is convex seen from below. The curve is very similar to that assumed in Fig. 3 (note that the horizontal axis is reversed in Fig. 8B).

It is expected in a case like this that resistant strains would be fitter in insecticidal environments, and susceptible in unstressed environments, as in Table 1. Contrary to this expectation Eggers-Schumacher (1983) found that three resistant clones were superior to four susceptible clones in both fecundity and developmental period. If the same applies in field situations the resistant strains will soon replace the unsusceptible ones.

Although the theory has been presented in terms of the general relationship between stress levels and their consequences for growth or mortality rate, these examples show that the theory can also be framed in terms of the concentration of a toxin, as in Fig. 7, or the density of damage caused by stresses in the tissue of organisms, or even in terms of the extent to which body fluids become diluted as in osmotic stress. Call the concentration of toxin or damage in body fluids C. Then we need to define the relationships between C and μ and g and the impact of metabolic investment on C. In a sense, this is equivalent to making more explicit the causal link between the application of a stress and its impact on μ and g.

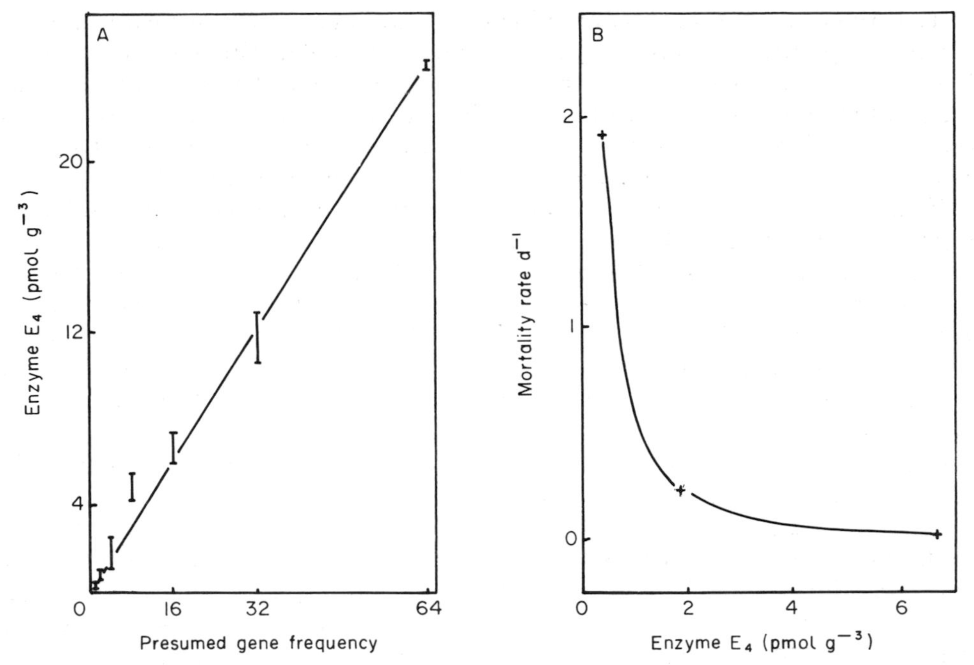

Figure 8. A, Concentration of the detoxifying enzyme E_4 in seven strains of the aphid *Myzus persicae* in relation to the number of copies of structural gene hypothetically present in each strain (from Oppenoorth, 1985, data from Devonshire & Sawicki, 1979). B, Daily mortality rate of three of the strains in relation to E_4 concentration. Aphids were placed on potato leaves that had been dipped in an organophosphorus insecticide (Demeton-S-methyl). We backcalculated mortality rate from LD_{50}s assuming for simplicity that survivorship was a negative exponental function of dose. Data from Sawicki & Rice (1978) and Devonshire & Sawicki (1979).

It seems plausible that the relationship between C and μ is nonlinear, with μ increasing increasingly with C. An example is the mortality effects of doses of X-irradiation (Traut, 1963; Perutz, 1987). Arguably Fig. 7C also shows this characteristic above low, physiological levels of C. On the other hand, it also seems likely that the relationship between investment in processes that reduce C (e.g. basal metabolism, call it R) is nonlinear, with the amount of reduction in C increasing with R, but at reducing rate, and never completely reducing to zero. Hence the relationship between C and g ($\propto$ (resource input) minus R) will be nonlinear, with C increasing increasingly with g, and therefore μ will also increase increasingly with g. The theory now becomes directly equivalent to that specified earlier.

WHAT THE THEORY HAS TO SAY ABOUT AGEING PROCESSES

It is possible to view ageing as a process involving the accumulation of cellular and molecular damage through the lifetime of an organism; i.e. due to intrinsic processes such as thermal noise, autoxidation, racemization, somatic mutation and errors in synthesis (Calow, 1978). Equally this damage can be repaired and/ or replaced but only at a metabolic cost (Kirkwood, 1987). Thus the density of damage (C) within the tissues can be understood in terms of the following balance:

$$C = \text{General damage} - \text{repair}.$$

In principle, therefore, repair can be set at a level that will hold C constant and lead to an immortal system! But this could only be achieved with a large investment in repair and hence a cost to production (both somatic and reproduction). What is the optimal investment? Here the theory is probably best cast in terms of adult mortality rate and reproductive investment (P_r) but the predictions are unaffected. Thus increased mortality stress will shift the $\mu(P_r)$ and the $\mu(g)$ curves up in the $P_r\mu$ and $g\mu$ planes and lead to an increased investment in P and g and a reduced investment in defence against ageing processes. In other words organisms exposed to high mortality risks, should show more direct signs of ageing once these risks are removed. Kirkwood (1987) using a similar model has come to the same conclusions.

Weak support for this theory is available from comparisons between different taxonomic groups (see below). The tests would be stronger if the shapes and positions of the trade-off curves were known, but data on these are lacking. We therefore assume for the purposes of these comparisons that, for given body weight in both taxonomic groups, (1) the same resources are available for metabolism; and (2) investment in defence results in the same reductions in mortality rate. If these assumptions are shown to be false, then the argument will need modification.

Comparing birds and mammals, it appears that birds, which on average live longer than ground-dwelling mammals of the same weight (see allometric relationships in Calder, 1984), perhaps because they are less subject to predation, also have longer maximum life spans (Comfort, 1979; Calder, 1984). This fits the theoretical prediction that organisms exposed to high mortality risk should show more signs of ageing once these risks are removed. Ostrich and emu, however, which are ground-dwelling birds, have similar maximum life spans to mammals of comparable weight, i.e. considerably less than expected from the allometric relationship for birds (Comfort, 1979; Calder, 1984). In evaluating food intake, birds and mammals are often treated together (see e.g. Calder, 1984), so resource intake may be the same (Assumption 1, above). If resource intake is the same, and if flying birds do spend more on defence against ageing, then there cannot be so many resources left for reproductive investment in flying birds. Figures for power allocation to reproduction are not available, but clutch/litter mass is marginally lower in birds, as predicted (Peters, 1983; Calder, 1984).

In general, well-defended organisms should be less subject to ageing processes than more poorly defended organisms—and hence the long lives of tortoises may not be apocryphal! Tortoise and turtle longevity is reviewed by Comfort (1979). By similar reasoning, poor growth conditions should select for less investment in ageing-resisting processes. Hence ageing should be more prevalent in low-productivity habitats. However, all the above arguments assume that the trade-off curves have the same shapes in the stressed and unstressed environments, but if this is not the case predictions could be reversed.

CONCLUSIONS

We have developed a general theory of stress. This gives a basis for a general classification of stresses as well as a framework for understanding the responses of organisms to them. However, whereas the theory is rigorous, and easy to apply if the shapes and positions of trade-off curves are known, discovering the latter is more of a challenge. Thus, the model makes certain assumptions about metabolic investments in stress resistance and the way that these trade-off against other aspects of metabolism. These metabolic processes and trade-offs, in turn, depend upon the mechanisms involved in stress resistance. So here the model invites further detailed investigations of these mechanisms and their implications for the metabolic trade-offs. Once these are established, the model makes a number of testable predictions about optimal responses to particular stresses. Here, the arguments have been framed as evolutionary responses to stresses acting as selection pressures. However, the models can equally be used to represent optimal, plastic responses of phenotypes to stresses acting as proximate influences on organisms—and these kinds of plastic responses and their evolution deserve further, careful attention. Finally, an important and attractive feature of the theory is its generality: it can address questions concerned with a variety of stresses, both extrinsic (pollutants, predators, radiation, osmotic stress etc.) and intrinsic (ageing processes) to the organism and it can represent a variety of stress-resisting mechanisms (e.g. predator defence, disease resistance, repair of damage caused by xenobiotics and radiation). Because of this, we hope that it will form a focus for the currently rather disparate studies that are carried out as stress biology.

ACKNOWLEDGEMENTS

Suggestions from G. Holloway, R. Aronson, R. H. Smith, A. R. McCaffery, S. P. Hopkin, A. L. Devonshire, D. M. Broom and K. Simkiss are gratefully acknowledged.

REFERENCES

BAYNE, B. L., MOORE, M. N., WIDDOWS, J., LIVINGSTONE, D. R. & SALKELD, P., 1979. Measurement of responses of individuals to environmental stress and pollution: studies with bivalve molluscs. *Philosophical Transactions of the Royal Society, London, 286B:* 563–581.

BENGTSSON, G., GUNNARSSON, T. & RUNDGREN, S., 1983. Growth changes caused by metal uptake in a population of *Onychiurus armatus* (Collembola) feeding on metal polluted fungi. *Oikos, 40:* 216–225.

BENGTSSON, G., GUNNARSSON, T. & RUNDGREN, S. 1985. Influence of metals on reproduction, mortality and population growth in *Onychiurus armatus* (Collembola). *Journal of Applied Ecology, 22:* 967–978.

BEVERTON, R. J. H. & HOLT, S. T., 1959. A review of lifespans and mortality rates of fish in nature, and their relation to growth and other physiological characteristics. *CIBA Foundation Colloquium on Ageing, 5:* 142–180.

BISHOP, J. A., 1981. A neoDarwinian approach to resistance: examples from mammals. In J. A. Bishop & L. M. Cook (Eds.), *Genetic Consequence of Man-Made Change:* 37–51. London: Academic Press.

BRADSHAW, A. D. & McNEILLY, T., 1981. *Evolution and Pollution. Studies in Biology, 130.* London: Edward Arnold.

CALDER, W. A., 1984. *Size, Function and Life History.* Cambridge, Massachusetts: Harvard University Press.

CALOW, P., 1978. Bidder's hypothesis revisited: a solution to some key problems associated with the general molecular theory of ageing. *Gerontology, 24:* 448–458.

COMFORT, A., 1979. *The Biology of Senescence.* Edinburgh and London: Churchill Livingstone.

CONNELL, J. H. & ORIANS, E., 1964. The ecological regulation of species diversity. *American Naturalist, 98:* 399–414.

COOK, S. C. A., LEFEBVRE, C. & McNEILLY, T., 1972. Competition between metal tolerant and normal plant populations on normal soil. *Evolution, 26:* 366–372.

DEVONSHIRE, A. L. & SAWICKI, R. M., 1979. Insecticide-resistant *Myzus persicae* as an example of evolution by gene duplication. *Nature, 280:* 140–141.

EBERT, T. A., 1982. Longevity, life history and relative body wall size in sea urchins. *Ecological Monographs, 52:* 353–394.

EBERT, T. A., 1985. Sensitivity of fitness to macroparameter changes: an analysis of survivorship and individual growth in sea urchin life histories. *Oecologia, 65:* 461–467.

EGGERS-SCHUMACHER, H. A., 1983. A comparison of the reproductive performance of insecticide-resistant and susceptible clones of *Myzus persicae*. *Entomologia Experimentalis et Applicata, 34:* 301–307.
GARLICK, P. J., 1980. Assessment of protein metabolism in the intact animal. In P. J. Buttery & D. B. Lindsay (Eds), *Protein Deposition in Animals:* 51–67. London: Butterworth.
GARLICK, P. J., BURK, T. L. & SWICK, R. W., 1976. Protein synthesis and RNA in tissues of the pig. *American Journal of Physiology, 230:* 1108–1112.
GRIME, J. P., 1979. *Plant Strategies and Vegetation Processes*. Chichester: John Wiley.
HAWKINS, A. J. S., 1985. Relationships between the synthesis and breakdown of protein, dietary absorption and turnovers of nitrogen and carbon in the blue mussel, *Mytilus edulis* (Bivalvia: Mollusca). *Oecologia, 66:* 42–49.
KIRKWOOD, T. B. L., 1987. Maintenance and repair processes in relation to senescence: adaptive strategies of neglect. In P. Calow (Ed.), *Evolutionary Physiological Ecology:* 53–66. Cambridge: Cambridge University Press.
KOEHN, R. K. & BAYNE, B. L., 1989. Towards a physiological and genetical understanding of the energetics of the stress response. *Biological Journal of the Linnean Society, 37:* 157–171.
LACK, D., 1968. *Ecological Adaptations for Breeding in Birds*. London: Methuen.
MEIER, P. R., PETERSON, R. G., BONDS, D. R., MESCHIA, G. & BATTALGIA, F. C., 1981. Rates of protein synthesis and turnover in fetal life. *American Journal of Physiology, 240:* 320–324.
MOORE, M. N., LIVINGSTON, D. R., WIDDOWS, J., LOWE, D. M. & PIPE, R. K., 1988. Molecular, cellular and physiological effects of oil-derived hydrocarbons on molluscs and their use in impact assessment. *Philosophical Transactions of the Royal Society, London, Series B, 316:* 603–623.
NICHOLAS, G. A., LOBLEY, G. E. & HARRIS, C. I., 1977. Use of the constant infusion technique for measuring rates of protein synthesis in the New Zealand White Rabbit. *British Journal of Nutrition, 38:* 1–17.
O'CONNOR, R. J., 1984. *The Growth and Development of Birds*. Chichester: John Wiley.
OPPENOORTH, F. J., 1985. Biochemistry and genetics of insecticide resistance. In G. A. Kerkut & L. I. Gilbert (Eds), *Comprehensive Insect Physiology, Biochemistry and Pharmacology, 12*. Oxford: Pergamon Press.
PERUTZ, M. F., 1987. Physics and the riddle of life. *Nature, 326:* 555–558.
PETERS, R. H., 1983. The ecological implications of body size. Cambridge: Cambridge University Press.
POTTS, W. T. W. & PARRY, G., 1964. *Osmotic and Ionic Regulation in Animals*. Oxford: Pergamon.
ROUSH, R. T. & McKENZIE, J. A., 1987. Ecological genetics of insecticide and acaricide resistance. *Annual Review of Entomology, 32:* 361–380.
SAWICKI, R. M. & RICE, A. D., 1978. Response of susceptible and resistant peach-potato aphid *Myzus persicae* (Sulz.) to insecticides in leaf-dip bioassays. *Pesticide Science, 9:* 513–516.
SIBLY, R. M., 1989. What evolution maximizes. *Functional Ecology,* in press.
SIBLY, R. M. & CALOW, P., 1984. Direct and absorption costing in the evolution of life cycles. *Journal of Theoretical Biology, 111:* 463–473.
SIBLY, R. M. & CALOW, P., 1986. *Physiological Ecology of Animals: an Evolutionary Approach*. Oxford: Blackwell.
SIBLY, R. M. & CALOW, P., 1987a. Growth and resource allocation. In P. Calow (Ed.), *Evolutionary Physiological Ecology*. Cambridge: Cambridge University Press.
SIBLY, R. M. & CALOW, P., 1987b. Ecological compensation—a complication for testing life-history theory. *Journal of Theoretical Biology, 125:* 177–186.
TERRIERE, L. C., 1984. Induction of detoxification enzymes in insects. *Annual Review of Entomology, 29:* 71–88.
TRAUT, H., 1963. Dose-dependence of the frequency of radiation-induced recessive sex-linked lethals in *Drosophila melanogaster*, with special consideration of the stage sensitivity of the irradiated germ cells. In F. H. Sobels (Ed.), *Repair from Genetic Radiation Damage*. London: Pergamon.
VERMEIJ, G. J., 1978. *Biogeography and Adaptation: Patterns of Marine Life*. Cambridge, Massachusetts: Harvard University Press.
VERMEIJ, G. J. & COVICH, A. P., 1978. Coevolution of freshwater gastropods and their predators. *American Naturalist, 112:* 833–843.
VERMEIJ, G. J. & DUDLEY, E. C., 1985. Distribution of adaptations: a comparison between the functional shell morphology of freshwater and marine pelecypods. In E. R. Trueman & M. R. Clarke (Eds.), *The Mollusca, Volume 10, Evolution*. Orlando: Academic Press.
WATERLOW, J. C., 1980. Protein turnover in the whole animal. *Invertebrate Cellular Pathology, 3:* 107–119.
WIDDOWS, J., DONKLIN, P., SALKELD, P. N., CLEARY, J. J., LOWE, D. M., EVANS, S. V. & THOMSON, P. E., 1984. Relative importance of environmental factors in determining physiological differences between two populations of mussels (*Mytilus edulis*). *Marine Ecology Progress Series, 17:* 33–47.
WILSON, J. B., 1988. The cost of heavy-metal tolerance: an example. *Evolution, 42:* 408–413.
WOOD, R. J. & BISHOP, J. A., 1981. Insecticide resistance: genes and mechanisms. In J. A. Bishop & L. M. Cook (Eds.), *Genetic Consequences of Man-Made Change:* 97–127. London: Academic Press.

Biological Journal of the Linnean Society (1989), *37:* 117–136.

An integrated approach to environmental stress tolerance and life-history variation: desiccation tolerance in *Drosophila*

ARY A. HOFFMANN

Department of Genetics and Human Variation, La Trobe University, Bundoora, Victoria 3083, Australia

AND

P. A. PARSONS

Division of Science and Technology, Griffith University, Nathan, Queensland 4111, Australia

The availability of metabolic energy provides a general measure of the environmental stress that can be tolerated by organisms, leading to the hypothesis that increased tolerance to a range of environmental stresses will be associated with a reduction in metabolic rate in *Drosophila* and many other organisms. This hypothesis makes three predictions about genetic variation for stress tolerance: (1) increased stress tolerance will tend to be associated with decreased metabolic rate; (2) genetic correlations between tolerance of different environmental stresses will tend to be positive; (3) stress tolerance and life-history traits will tend to be genetically correlated; in *Drosophila* correlations with life-history traits other than longevity will tend to be negative.

These predictions were tested by artificially selecting for increased desiccation tolerance in *Drosophila melanogaster,* using an 85% mortality level. The response to selection was rapid and the mean realized heritability was *c.* 0.65. The selection response was associated with a decreased rate of water loss, reduced activity and a decrease in metabolic rate in agreement with prediction (1). Selection did not alter body size. Selected lines were relatively more tolerant of starvation and a toxic concentration of ethanol in agreement with prediction (2), and had lower fecundities in agreement with prediction (3).

KEY WORDS:—Environmental stress – metabolic rate – tolerance – *Drosophila* – desiccation.

CONTENTS

0024–4066/89/050117+20 $03.00/0

INTRODUCTION

Environmental stress and evolutionary change

Environmental stress plays a major role in determining the distribution and abundance of many organisms. Two examples are the importance of extreme temperatures in explaining the distribution of Australian floral types (Nix, 1981; Webb, Tracey & Williams, 1984) and of Australian *Drosophila* species (Parsons, 1981). Extreme stresses which tend to be irregular in their occurrence can impose a high selection intensity and lead to rapid phenotypic changes in populations at stressful times. For example, in the finch *Geospiza fortis* in the Galapagos Islands, only 15% of the birds survived a major drought period in 1977 (Boag & Grant, 1981). During this period, birds were selected for increased beak depth and body weight, and decreased beak width. A less extreme stress occurred in the classic study of Bumpus (1899), where about 50% of a sparrow population survived a winter storm. A reanalysis of the data (Lande & Arnold, 1983) indicated that small birds were favoured. A major difficulty in studying stress periods is their unpredictability, so that observations over a long period of time are necessary. For this reason, evolutionary biologists have tended to underestimate their importance (Parsons, 1987a).

Genetic variation for tolerance to extreme temperatures and desiccation stress has been demonstrated in several studies, particularly those on *Drosophila melanogaster* using isofemale strains set up at random from natural populations (Parsons, 1980a). There is also substantial genetic variability for responses to other stresses in natural populations, often with genes or gene complexes with differing but rather large additive effects (Parsons, 1974), for example anoxia in *D. melanogaster* measured by sensitivity to CO_2 (Matheson & Parsons, 1973). Tolerance to NaCl in *D. melanogaster* can be increased rapidly by continuous directional selection under laboratory conditions (Waddington, 1959; Miyoshi, 1961). In *Geranium carolinianum,* a heritability (narrow sense) of 0.54 ± 0.14 was obtained for SO_2 resistance (Taylor, 1978) indicating substantial additive genetic variability and hence the potential for rapid adaptation to high SO_2 conditions (Horsman, Roberts & Bradshaw, 1979).

Evidence from an array of organisms indicates that both phenotypic and genotypic variability tend to be high under conditions of severe stress imposed by the physical and biotic environments (Parsons, 1987a), provided, of course, that the stress is not so severe as to cause complete lethality. An early but convincing example is in the crucifer, *Arabidopsis thaliana*, where the genotypic variance for growth rate at the extreme temperature of 31.5°C was more than five times that at lower more optimal temperatures (Langridge & Griffing, 1959). These genetic considerations imply that rapid genetic changes may occur under stressful conditions, which are of short duration relative to the total time under consideration.

Mechanisms underlying adaptation to stress

At the biochemical level, a general way that the stress experienced by organisms can be assessed is via the availability of metabolic energy (Ivanovici & Wiebe, 1981). This is measured as the adenylate energy charge (AEC) which is an index calculated from measured amounts of adenosine triphosphate (ATP), adenosine diphosphate (ADP) and adenosine monophosphate (AMP) whereby (Atkinson, 1977; Hochachka & Somero, 1984)

$$\text{AEC} = (\text{ATP}+\tfrac{1}{2}\text{ADP})/(\text{ATP}+\text{ADP}+\text{AMP}).$$

Values between 0.8 and 0.9 are typical of organisms where environmental conditions are optimal or non-stressed and there is often active growth and reproduction. In organisms under recognizable limiting or non-optimal conditions, AECs are typically in the range 0.5 to 0.7, and under severe stress where viability losses become apparent even following a return to normal conditions AECs are in the 0.5 region (Ivanovici & Wiebe, 1981).

Individuals that are genetically tolerant to environmental stress are expected to have higher AECs when they are exposed to stressful conditions than individuals that are genetically sensitive. This prediction applies to most types of environmental stress because AECs have been found to change in response to a wide range of environmental perturbations including nutritional stress, oxygen depletion, desiccation, heat and chemical stresses (Ivanovici & Wiebe, 1981).

One way that an organism may increase its tolerance to a range of environmental stresses is to reduce its metabolic energy requirement via a reduction in metabolic rate. Such a mechanism is used by animals that hibernate or undergo diapause to avoid periods of environmental stress. An extreme case is represented by diapausing insects in which metabolic rate can be up to 2000 times lower than their normal metabolic rate (Hochachka & Somero, 1984). Stress tolerance is also known to be lower under conditions that increase metabolic rate. For example, metabolic rate increases over the non-lethal temperature range in *D. melanogaster* and in other insects (Hunter, 1964), and adults show a reduction in tolerance to a range of stresses at higher temperatures, including anoxia, starvation, high concentrations of ethanol, and desiccation (Matheson & Parsons, 1973; Hoffmann & Parsons, unpublished).

These considerations lead to two predictions about genetic variation for tolerance to many environmental stresses. First, individuals that are genetically tolerant to stress should have a reduced metabolic rate. Second, tolerance to different environmental stresses should be positively correlated genetically because genes decreasing metabolic rate are likely to increase tolerance to a number of stresses. These predictions will not apply to all cases of stress tolerance, particularly those associated with chemical stresses. For example, heavy-metal tolerance in grasses is usually specific to a particular metal (Turner,

1969) and resistance to insecticides is often associated with a specific detoxification mechanism (Plapp, 1976). Tolerance could even be associated with increased metabolic rate in these cases because of an increased rate of detoxification. Genetic correlations may also be modified by factors other than gene action, in particular the history of selection in a population (Falconer, 1981). For example, joint continuous selection for increased tolerance of two stresses may result in the rapid fixation of genes contributing to increased tolerance of both stresses. Much of the remaining genetic variation may be restricted to genes that have positive effects on one trait and negative effects on another trait, diminishing the positive correlation between the stress tolerance traits.

There is some published evidence for an association between genetic variation in metabolic rates and tolerance of environmental stresses. In *D. melanogaster*, there is indirect evidence that desiccation-resistant isofemale strains have a lower metabolic rate as measured by oxygen consumption than do desiccation-sensitive strains (Parsons, 1974). Lines of *D. melanogaster* selected for postponed senescence were more tolerant of environmental stress and had reduced metabolic rates and higher lipid contents compared to control lines (Service, 1987). Finally the basal metabolic rate of four karyotypes of the mole rat *Spalax ehrenbergi* vary with climatic conditions, decreasing towards the desert where conditions are relatively more stressful in terms of heat and desiccation (Nevo & Shkolnik, 1974).

A few studies suggest the predicted positive genetic correlations between tolerance to different environmental stresses. Adult *D. melanogaster* populations selected for postponed senescence were more resistant than control flies to desiccation, starvation and a toxic concentration of ethanol (Service *et al.*, 1985). Extensive selection experiments for genetic adaptation to temperature extremes have been carried out in *Aphytis lingnanensis,* a hymenopterous parasite of the Californian red scale, *Aonidiella aurantii,* using a heterogeneous gene pool made up of a combination of strains from Formosa, China, Texas and Mexico (White, DeBach & Garber, 1970). In the low-temperature line there was a small correlated increase in tolerance to high temperature. There is limited evidence for an association between mortalities under extreme heat stress and high doses of Co^{60}–γ irradiation (Ogaki & Nakashima-Tanaka, 1966; Parsons, 1969; Westerman & Parsons, 1973).

Unfortunately, few experiments have been carried out under extreme stresses where major genes are likely to be apparent and where lethality is close (AECs in the vicinity of 0.5). Experimental studies in evolutionary biology tend to use more optimal environments (Parsons, 1987a). Large additive genetic effects are more likely to predominate at extreme stress levels. For example, high doses (1–1.2 kGy) of Co^{60}–γ irradiation give high levels of additivity enabling the localization of genetic activity to chromosomal regions, whereas non-additivity predominates at lower doses (0.4–0.8 kGy) (Parsons, MacBean & Lee, 1969; Westerman & Parsons, 1973). A high temperature of 29.5°C reduces longevity to that found between 0.6–0.8 kGy in the irradiation experiments, and in parallel the genetic architecture was predominantly non-additive.

Environmental stress and life-history traits

A large number of genetic studies of life-history traits have recently been carried out, using relatively simple forms of genetic analysis. Heritabilities of life-history characteristics tend to be relatively low, with typical values of the order of 0.2–0.3 (Rose, 1983; Charlesworth, 1984; Roff & Mousseau, 1987), and this complicates genetic analysis of these traits. In contrast, ecological phenotypes such as resistance to extreme stresses show higher genotypic components, especially when stresses are close to lethality.

One important retrospective survey (Nevo, Beiles & Ben-Shlomo, 1984), where life-history characteristics have been studied simultaneously with stress traits, concerns an analysis of the genetic diversity of enzymes in a wide range of species in relation to ecological parameters (life zone, geographic range, habitat type and range, climatic region), demographic factors (species size, population structure, gene flow, sociality) and a series of life-history characteristics (longevity, generation length, fecundity, origin, and parameters relating to the mating system and mode of reproduction). For a range of populations, species and higher taxa, variation in proteins was found to be best accounted for by ecological parameters, many of which indirectly incorporate climatic extremes. For example, in the genus *Drosophila* about 66% of genetic diversity could be explained by ecological factors and 23% by life-history factors. This emphasizes (1) the more direct relationship between ecological phenotype and habitat compared with life-history traits, and (2) the need for simultaneous studies of life history and ecological phenotypes in the same population under optimal and stressful environments (Parsons, 1987b).

Variation in many life-history traits is expected to be correlated with the availability of metabolic energy. Case studies of selection at single enzyme loci indicate that genotypes that maximize flux through biochemical pathways are favoured by selection (Watt, 1985). Genetic differences in rates of flight metabolism between isogenic chromosome lines of *D. melanogaster* show a correlation of > 0.9 with the mechanical power output of flight muscles, a measure of flight ability (Laurie-Ahlberg *et al.*, 1985). This trait is related to reproductive success because it will affect the ability of flies to locate food and obtain mates. Development time is likely to be decreased by an increase in metabolic rate: variation in the activity of six out of seven enzymes are correlated with pre-adult development rate in *D. melanogaster* and *D. subobscura* (Marinkovic, Milosevic & Milanovic, 1986; Cluster *et al.*, 1987).

Increased fitness for such life-history traits is positively correlated with the availability of metabolic energy, which may lead to a positive correlation with metabolic rate. In addition, mechanisms that increase stress tolerance may divert energy and other resources away from growth and reproduction under optimal conditions. This leads to the expectation that tolerance of environmental stresses in *Drosophila* may show a negative genetic correlation with many life-history traits. Service *et al.* (1985) and Service & Rose (1985) found such a negative correlation between early fecundity and tolerance of three environmental stresses (desiccation, ethanol, nutrition) in *D. melanogaster.* Other life-history characteristics can be similarly considered according to metabolic rate arguments: for example increased development time will presumably be associated with low metabolic rate and high resistance to stress. Conversely increased mating ability, fecundity and fertility may be correlated with low resistance to stress.

The expected negative correlations may be modified when life-history traits are measured under stressful conditions. For example, Service & Rose (1985) found a negative correlation of −0.91 between fecundity and starvation under a standard laboratory environment. Under a novel, and hence somewhat stressful environment consisting of a low temperature (15.5°C) and culture in the dark, the negative correlation became significantly lower at −0.45. As conditions became more stressful, correlations may be biased towards positive values because life-history traits will be measuring some component of stress tolerance.

Increased fitness for life-history traits may also be achieved by changes in metabolic efficiency rather than metabolic rate. Increased metabolic efficiency can decrease oxygen demand, particularly in organisms living in environments where oxygen is limiting. For example, degree of heterozygosity is positively correlated with growth rate and negatively correlated with routine oxygen consumption in aquatic organisms such as oysters (Koehn & Shumway, 1982) and rainbow trout (Danzmann, Ferguson & Allendorf, 1988). In tiger salamanders, heterozygosity is also positively correlated with growth rate and negatively correlated with standard oxygen consumption involving maintenance functions; however, heterozygosity is positively correlated with oxygen consumption during activity (Mitton, Carey & Kocher, 1986). These examples suggest that it will be difficult to make general predictions about correlations between metabolic rate and life history traits unless factors such as habitat and the past history of selection in a population are taken into account (Southwood, 1988).

Predictions

The above discussion on environmental stresses and life-history traits leads to predictions about genetic variation for these traits in *Drosophila*.

1. Stress-tolerant genotypes will tend to have lower metabolic rates.
2. Genetic correlations between many stress traits will tend to be positive, especially under stress levels where lethality is close.
3. Genetic correlations between many stress traits and life-history traits may tend to be negative regardless of stress, except for longevity and associated traits because stress-tolerant genotypes may tend to live longer.

These predictions need to be tested in populations showing genetic heterogeneity for tolerance to environmental stresses, based upon hypotheses that can be developed from AEC and metabolic rates. This would involve investigating metabolic rate and life-history traits in natural populations that have diverged genetically because of exposure to different levels of stress. The predictions can also be tested with laboratory strains that have been selected for increased or decreased tolerance of an environmental stress.

Desiccation tolerance in Drosophila *as a case study*

Here we consider desiccation tolerance in *Drosophila melanogaster*, concentrating on its response to selection and on correlated changes in morphological traits, physiological traits, fitness components and tolerance to other environmental stresses. Work in *D. melanogaster* on desiccation tolerance has demonstrated substantial variation among isofemale strains and between geographic localities (Parsons, 1970, 1980b). Populations from habitats where desiccation stress is likely to be sporadically severe tended to be more resistant than populations from more benign habitats. Variation among isofemale strains tended to be lower in populations from stressed habitats, which is consistent with more intense selection for desiccation stress. The ecological significance of the trait is also evident from interspecific experiments which show that desiccation tolerance correlates with habitat. For example, temperate-zone *Drosophila* species are far more resistant to desiccation than tropical-zone species (Parsons, 1981). Associations between desiccation tolerance and habitat have also been found in *Aedes aegypti* and *A. atropalpus* (Machado-Allison & Craig, 1972).

The physiological basis of desiccation tolerance in insects has been extensively studied, but little is known about its genetic basis. Factors affecting the desiccation tolerance of mutants at the adipose locus of *D. melanogaster* have been identified (Clark & Doane, 1983). The investigation of the genetic and physiological basis of naturally occuring variation in *Drosophila* and other insects has been neglected, although as already stated desiccation-tolerant isofemale strains of *D. melanogaster* apparently have a lower metabolic rate than do desiccation-sensitive strains (Parsons, 1974).

We describe preliminary experiments based upon lines selected for desiccation tolerance with the following specific aims.

1. To determine the nature and extent of response at an extreme stress level. On the basis of earlier data we predict that the response to selection will be rapid.

2. To examine the morphological and physiological basis of the genetic response to desiccation stress. We predict that metabolic rate accounts at least partly for the response to selection.

3. To test for positive genetic correlations with other environmental stresses. We consider tolerance of starvation and tolerance of a high ethanol concentration that reduces longevity.

4. To test for negative genetic correlations between desiccation tolerance and life-history traits. We consider early female fecundity.

MATERIALS AND METHODS

Stocks

Selection was carried out on a mass-bred *D. melanogaster* stock initiated with the progeny of 30 inseminated females. The females were collected by fruit baiting in suburban Melbourne, Australia, in October, 1986. This stock has been maintained in the laboratory for six generations by mass-transfer of about 300 adults prior to selection. Flies were cultured at 19–21°C on a medium containing sucrose (9%), agar (3%) and dead yeast (6%), with nipagin and propionic acid as preservatives. Flies were maintained either in 600 ml bottles or in 40 ml vials.

Selection

As in *Geospiza fortis* (Boag & Grant, 1981), an intense level of selection giving 85% mortality was used. Three selection lines and three unselected control lines were independently maintained. The population size of each line was kept at 100 females (see below). An additional control population (line C) was maintained independently from the unselected lines. This was established with 480 flies at a density of 40 adults per bottle. Adults were repeatedly transferred to new bottles at two-day intervals so that flies from line C were always available. This stock was used to correct for day-to-day fluctuations in the selection response (see below).

Flies to be selected were collected from bottles when they were 0–3 days old. Flies were aged (sexes mixed) in vials with laboratory medium for 3–4 days at a density of 30–50 flies per vial. Females were then separated from males under CO_2 anaesthesia, and held at a density of 20 per vial for another 1–2 days. Only females were selected for desiccation tolerance. Flies were aged at the temperature (24–26°C) used for desiccation. Flies were desiccated in empty vials covered with gauze in a glass desiccator with silica gel as the desiccant.

To correct for day-to-day variation in conditions for desiccation, ten vials with females from the C line were included in each desiccator. The time taken for 50% of the females to die in these vials was scored by counting the number of dead individuals at one-hour intervals. The 50% mark was linearly interpolated when more than half the flies had died.

Flies were removed from the desiccator when it was judged that 85% of them had died. Surviving females were counted and transferred by aspiration into bottles with 70–90 ml of laboratory medium and live yeast. Three bottles were set up per line, at a density of 33–36 females per bottle. This ensured a census size of 100 females per line. Preliminary experiments had indicated that most (> 90%) females were fertile after desiccation. The exact selection intensity was determined from the number of females that died. Flies from unselected lines were treated similarly except that they were not desiccated. Adults were transferred to fresh bottles several times to ensure sufficient numbers for the next generation of selection.

Comparing selected and control lines

The desiccation tolerance of the selected and unselected lines was compared twice during the course of selection. Two bottles were set up per line at a density of 20 males and 20 females per bottle. Adults from the selected lines had not been desiccated. Flies were cleared after one day so that larvae developed in uncrowded conditions. Progeny from different culture bottles were collected and tested separately. Females were aged and handled as during selection.

Three vials of females were obtained from each culture bottle. Replicate vials were assigned to different desiccators, so that the experimental design consisted of three randomized blocks. The design is given by the equation

$$Y_{ijklm} = a + b_i + c_{j(i)} + d_{k(ij)} + e_i + f_{m(ijkl)} \quad (1)$$

where a is the grand mean, b_i the selection term, $C_{j(i)}$ the replicate line term nested within selection, $d_{k(ij)}$ the bottle term nested in selection and replicate line, e_i the desiccator (block) term, and $f_{m(ijkl)}$ the error term.

Vials were placed along the periphery of the desiccator so that the number of dead flies could be counted without removing the desiccator lid. Counts were made at one-hour intervals until at least half the flies had died, and the LT_{50} was linearly interpolated for each vial. In the comparison of the selected and control lines after nine generations of selection, LT_{85}s were also scored by monitoring the vials until at least 85% of the flies had died.

Morphological changes

Females (5–6 days old) were obtained from lines after nine generations of selection. They were initially weighed on a microbalance accurate to ± 0.02 mg directly after being etherized. Females were then mounted on microscope slides for wing removal. Head width was measured under a binocular microscope, using the widest point from eye to eye. Wing length was measured from the intersection of the anterior cross vein and longitudinal vein 3 (L3) to the intersection of L3 with the distal wing margin. Wing width was measured from the intersection of L5 and the wing margin to the intersection of L2 and the wing margin.

Flies from lines were weighed and measured in a randomized order. Twenty flies were measured per line, ten from each of two culture bottles. The experimental design is given by

$$Y_{ijklm} = a + b_i + c_{j(i)} + d_{k(ij)} + f_{m(ijkl)} \tag{2}$$

which is similar to the experimental design in (1) except for the absence of a block effect.

Weight loss after three hours

Weight loss due to short-term desiccation was measured by placing groups of 20 live females in a desiccator for three hours. Groups were weighed to the nearest 0.1 mg before and after desiccation. Weight loss of groups of dead flies was similarly determined except that flies were etherized directly after they were intially weighed. The experimental design followed (1). Three replicate groups of females were tested from each of two culture bottles. Flies from the same replicate were weighed as one batch and placed together in a desiccator. Weight loss was expressed as the percentage of the initial wet weight lost.

Mortality and weight changes of individual females

The desiccation tolerance of females not held in groups was examined by placing single females into small plastic tubes (length 30 mm, internal diameter 4 mm) sealed at one end and covered with gauze at the other end. The tubes were taped to the side of a desiccator, and mortality was scored at one-hour intervals. The experimental design followed (2).

Weight loss at death and total water content were obtained from the wet weight, dry weight and weight at death of individual flies. Females were initially weighed and placed into tubes. Mortality was monitored at 30-min intervals. Females were weighed immediately after they died, and again after they had been dried for five days at 37°C. The difference between wet and dry weights provides an estimate of the water content of the female, and the difference between wet and dead weights provides an estimate of the weight loss flies can tolerate before dying. These measures were expressed as percentage of the initial wet weight.

Activity and metabolic rate

To measure activity of flies in the desiccator, paper strips marked with lines 10 mm apart were taped to the back of vials. Twenty females were placed into each vial as for selection, and the vials were arranged on the periphery of a desiccator. Vials were scored for activity on three occasions, three, four and five hours after vials had been placed into the desiccator. On each occasion, the five females in a vial that were nearest the middle line on a strip were watched for 30 seconds. The number of females that moved were counted. Counts for a vial were summed over the three observation periods so that activity was expressed as the number of females that moved out of a maximum of 15. The experimental design follows (1) in that replicate vials were placed in different desiccators. Activity was also examined under high humidity conditions. Vials were placed in desiccators containing water instead of silica gel (i.e. hydration chambers).

Metabolic rate of the lines was determined in a Gilson differential respirometer. Flies were held in the same type of plastic tube that was used for the desiccation of individual females (see above). Ten females from a line were aspirated into a tube and two tubes were placed into a flask which was immersed in a 25°C water bath. A 10% KOH solution was used to remove CO_2. Oxygen consumption was measured over 2.5 hours. Twelve flasks were set up simultaneously. The experimental design followed (1) in that a block effect was associated with different runs of the respirometer.

Tolerance of starvation and ethanol

To examine tolerance of starvation under humidity, vials with females were placed in a hydration chamber with distilled water. The number of dead flies in a vial was scored at 12-hour intervals until at least half the females had died. The experimental design followed (2) because all replicates were placed in the same hydration chamber.

Tolerance of a high concentration of ethanol (15%) was measured following the procedure of Starmer, Heed & Rockwood-Sluss (1977). Ten ml of a 15% (v/v) solution of ethanol was pipetted into a vial with approximately 1 g of cotton wool. The vial was covered with gauze before a second vial with females was inverted over it. The two vials were then hermetically sealed with laboratory film. Mortality of the females was scored at 6–10 hour intervals. The experimental design followed (2) with six replicates per line.

Fecundity

Flies for this experiment were cultured under low-density conditions in vials. One male and one female were placed in a vial with yeasted medium and discarded after two days. Thirty vials were set up per line. Early fecundity of the progeny was scored following the method of Rose & Charlesworth (1981). Females and males (0–1 day old) were kept as pairs in yeasted vials for two days for mating, and pairs were then transferred to laying vials for 24 hours. The latter contained 10 ml of laboratory medium covered with a yeast suspension. Flies were transferred to fresh laying vials for two additional 24–hour periods. Eggs in these vials were counted and the cumulative count was used as an estimate of early fecundity. The position of the vials was randomized according to line and the experimental design followed (1), except that there was no bottle effect. This experiment was carried out at 24°C.

RESULTS

Response to selection

The response was rapid despite selection being carried out on only one sex. All three selected lines had diverged from control lines after four generations of selection (Table 1), with a mean difference in desiccation tolerance of 4.0 hours (based on LT_{50}s). This difference increased to a mean of 10.0 hours (or more than 50% of the tolerance of the control lines) after nine generations. Selection accounts for 88% of the variance in the ANOVA on the generation nine data. There was no evidence for divergence among the replicate lines or for an effect of culture bottle on desiccation tolerance.

The selection response estimated from LT_{85}s was similar to the response estimated from LT_{50}s (Table 1); the mean difference between the selected and control lines was 9.2 hours. The similarity in the selection response as measured by LT_{50}s and LT_{85}s suggests that the rate of death between the 50% and 85% mortality points was not altered by selection. This was tested directly by calculating the time between LT_{50} and LT_{85} for each vial (data not presented). An ANOVA indicated that the selection effect was not significant ($F_{(1,30)} = 0.88$ when the mean square for the denominator consists of pooled replicate line, bottle and error terms) confirming that the mortality points were not altered.

The selection response was also examined by estimating genetic gain. The selection intensity was calculated each generation from female mortality. Genetic gain was estimated from a regression of the time flies took to reach 85% mortality on the cumulative selection intensity. The method of Muir (1986) and Hoffmann & Cohan (1987) was used to correct for day to day fluctuations in desiccation tolerance as reflected in the LT_{50}s of line C. The genetic gain of each line was estimated with multiple regression. The selected phenotype was initially modelled as a polynomial function of the cumulative selection intensity (i) and the control phenotype (c): $Y = a + b_1 i + b_2 i^2 + b_3 c + E$, where a is the intercept,

TABLE 1. Differences between selected and control lines tested after four and nine generations selection. Means and standard deviations are based on six replicates of 20 flies

		Generation 4	Generation 9	
		LT_{50}	LT_{50}	LT_{85}
Mean time in hours (S.D.)				
Selected lines	1	25.6 (2.5)	27.0 (1.4)	32.7 (1.5)
	2	26.2 (1.1)	28.4 (2.2)	34.8 (2.4)
	3	26.7 (1.8)	30.2 (2.6)	35.2 (2.2)
Control lines	1	22.9 (2.7)	17.8 (3.7)	25.0 (1.4)
	2	22.2 (4.5)	20.1 (4.1)	25.9 (0.9)
	3	21.5 (4.3)	17.8 (2.7)	24.2 (1.4)
Mean squares for ANOVAs				
Selection (d.f. = 1)		277.9***	901.0***	762.1***
Replicate line (d.f. = 4)		5.4	13.1	7.5
Bottle (d.f. = 6)		6.3	12.5	4.8
Desiccator (d.f. = 2)		19.2	4.2	5.6
Error (d.f. = 22)		9.6	8.1	5.1

***$P < 0.001$.

b_js are the partial regression coefficients and E is the error term. The quadratic term for i was initially included to allow for attenuation of the selection response, but this was excluded in the final regression equations of all lines because it was not significant ($P > 0.25$). Genetic gain (per unit of selection intensity) in the resultant equations is given by b_1. This method is superior to the usual approach of subtracting the control phenotype from the selected phenotype for correcting fluctuations because it corrects for any trends in the control during selection.

Estimates of genetic gain and the cumulative selection intensity are given in Table 2. The predicted increases in tolerance after selection are similar to the differences between selection and control lines from Table 1 when an adjustment is made for the additional generation of selection included in the predicted increases. Genetic gains do not represent heritabilities because desiccation tolerance was measured on groups of females. However, these estimates can be related to heritability by the phenotypic standard deviation of desiccation tolerance of individual females because the selection intensity multiplied by this standard deviation gives the selection differential (Falconer, 1981). Ten groups of 20 females from an unselected line were therefore desiccated in vials and mortality was scored at 30-minute intervals. The mean phenotypic standard deviation of female mortality in a vial was 2.50. Genetic gains of the three replicate lines were divided by this estimate and then multiplied by two (because selection was only carried out on one sex) to give realized heritability estimates of 0.69, 0.65 and 0.59.

Morphological changes

Three body measurements (head width, wing length, wing width) and wet weight were used to test for morphological changes associated with the selection response. Neither the body measurements nor wet weight showed a significant effect of selection (Table 3). There were no consistent differences between the means of the selected and control lines, and no evidence of divergence between the replicate lines. The only significant term in the ANOVAs was a culture bottle effect on fresh weight. Therefore selection did not alter head width or wing morphology and did not increase body size.

Weight loss after three hours

Females were desiccated for three hours to test whether the ability of flies to hold water contributed to the selection response. Live females from selected lines

TABLE 2. Response to selection: cumulative selection intensity, predicted response to selection, and genetic gain for the selected lines

	Line 1	Line 2	Line 3
Cumulative selection intensity	13.59	13.20	13.56
Predicted increase in tolerance after selection (hrs)[a]	11.81	10.80	10.02
Genetic gain	0.869	0.818	0.739

[a]Predicted increase calculated from genetic gain and the cumulative selection intensities.

TABLE 3. Morphological measurements made after nine generations of selection. Means and standard deviations are based on nine or ten flies

		Head width (mm)	Wing width (mm)	Wing length (mm)	Wet weight ($g \times 10^{-3}$)
Means (S.D.)					
Selected lines	1	0.91 (0.03)	1.17 (0.03)	1.73 (0.05)	1.52 (0.20)
	2	0.90 (0.04)	1.18 (0.03)	1.75 (0.05)	1.61 (0.19)
	3	0.93 (0.02)	1.18 (0.04)	1.76 (0.04)	1.63 (0.17)
Control lines	1	0.93 (0.05)	1.20 (0.05)	1.77 (0.05)	1.49 (0.26)
	2	0.93 (0.03)	1.18 (0.03)	1.73 (0.03)	1.57 (0.17)
	3	0.90 (0.03)	1.19 (0.03)	1.73 (0.05)	1.52 (0.18)
Mean squares for ANOVAs ($\times 10^{-3}$)					
Selection (d.f. = 1)		0.07	3.03	0.32	2.73
Replicate line (d.f. = 4)		1.99	1.09	3.56	1.86
Bottle (d.f. = 6)		0.57	0.81	0.65	7.17*
Error (d.f. = 47)		1.37	1.36	2.46	3.38

*$P<0.05$.

tended to lose less weight than females from control lines (Table 4). The selection term has borderline significance. Because the replicate selection lines had similar desiccation tolerance after selection and there were no culture-bottle effects on desiccation tolerance (Table 1), we decided to test if we could pool the mean squares of these effects with the error mean square (by the $P > 0.25$ criterion, Sokal & Rohlf, 1981) to increase the degrees of freedom for the denominator. There was a significant bottle effect in this experiment so that the bottle and error terms could not be pooled. However the selection term was significant when tested over pooled bottle and replicate line terms. The variance component due to selection accounts for 42% of the total variance. Means for all

TABLE 4. Percentage of wet weight lost by females after three hours of desiccation. Means and standard deviations are based on six replicates of 20 flies

		Live flies	Dead flies
Mean % of wet weight lost (S.D.)			
Selected lines	1	5.0 (0.8)	14.6 (5.3)
	2	3.9 (0.7)	9.9 (2.2)
	3	4.9 (1.8)	13.0 (4.0)
Control lines	1	6.5 (0.4)	13.5 (7.2)
	2	5.2 (0.9)	12.1 (6.7)
	3	7.8 (2.6)	22.7 (6.2)
Mean squares for ANOVAs[a]			
Selection (d.f. = 1)		2.93[b]	12.28
Repl. line (d.f. = 4)		0.62	12.12
Bottle (d.f. = 6)		0.52**	3.52
Block (d.f. = 2)		0.14	4.64
Error (d.f.)		0.12 (21)	2.93 (22)

[a]ANOVAs were carried out on arcsin transformed proportions ($\times 100$); [b]Selection term had borderline significance ($P<0.10$) when tested over replicate line and was significant ($P<0.05$) when tested over pooled error and replicate line terms; **$P<0.01$.

three control lines are higher than those from the selected lines, with an average difference of 1.9%, which is 29% of the weight loss in control lines.

Differences in the rate of weight loss may be associated with water loss through the cuticle or through the spiracles. To test for a cuticular component, females were first anaesthetized until death. This treatment leaves the spiracles open (Fairbanks & Burch, 1970). Dead females lost more than twice as much weight as live flies (Table 4). There was no effect of selection, suggesting that the higher desiccation tolerance of the selected lines was not associated with changes in the rate of water loss through the cuticle. The reduced weight loss in live flies was therefore probably associated with spiracle closure.

Mortality and weight changes of individual flies

Flies were desiccated individually in small containers to investigate if the difference in desiccation tolerance between selected and control lines depended on flies being held in groups. The selection term was highly significant when tested over the replicate line term (Table 5), accounting for 67% of the total variance. Females from the selected lines died an average of 7.0 hours after those from control lines, which represents 33% of the desiccation tolerance of the control lines. The difference between selected and control lines is smaller than line differences for females desiccated in groups (cf. Table 1).

To test the possibility that flies from selected lines tolerate a relatively higher level of water loss before dying, females were weighed directly after they died. Females had lost an average of 36% of their weight at the time of death (Table 5). There were no significant differences between the selected and control lines.

To examine variation in water content, individuals were dried until they stopped losing weight. The average water content of females was 73% (Table 5). Differences between selected and control lines were not significant. Selection also had no effect on the dry weight of individual flies (Table 5), in agreement with the absence of a correlated selection response for wet weight (Table 3).

TABLE 5. Mortality and weight loss of individual females. Means are based on 8–10 replicates

		Time until death (hours)	% Of initial weight lost		Dry weight ($g \times 10^{-3}$)
			At time of death	After drying	
Means (S.D.)					
Selected lines	1	29.9 (6.8)	38.4 (5.0)	73.9 (2.3)	4.3 (0.4)
	2	26.1 (4.2)	35.6 (5.9)	74.2 (2.0)	4.3 (0.5)
	3	28.1 (3.6)	34.7 (2.5)	72.8 (2.2)	3.9 (0.5)
Control lines	1	21.3 (2.4)	36.2 (5.5)	73.6 (2.4)	4.0 (0.4)
	2	21.9 (2.5)	35.1 (4.8)	72.5 (4.2)	4.2 (0.5)
	3	19.9 (2.6)	36.9 (4.9)	71.2 (3.5)	4.1 (0.4)
Mean squares for ANOVAs[a]					
Selection (d.f. = 1)		735.0**	0.16	4.33	0.054
Replicate line (d.f. = 4)		23.3	2.31	1.87	0.256
Bottle (d.f. = 6)		6.9	1.86	1.42	0.347
Error (d.f.)		16.9 (48)	2.93 (43)	1.86 (43)	0.166 (43)

[a]ANOVAs for % weight loss were carried out on arcsin transformed proportions (× 100); **$P<0.01$.

Activity and metabolic rate

Females from selected lines were less active than those from control lines when placed in a desiccator (Table 6). On average, females from the selected lines were 18% less likely to move than females from the control lines. The significant selection term accounted for 41% of the total variance. Females from selected lines were also relatively less active in a hydration chamber (Table 6). The selection term was significant in the ANOVAs, accounting for 55% of the total variance. Hence the activity difference between the lines does not depend on the presence of dry conditions.

Females from control lines consumed more oxygen than females from selected lines (Table 6) indicating that they had a relatively higher metabolic rate. There were no significant effects of bottle or replicate line in the ANOVA. The selection term is significant when tested over the replicate line term and the variance component due to selection accounts for 44% of the total variance.

Tolerance of starvation and ethanol

To examine the association between desiccation tolerance and tolerance of nutritional stress, females were kept without food under high humidity. Line means (Table 7) indicate that flies took much longer to die in the absence of a desiccation stress (cf. Table 1). The selection term was significant, and accounted for 72% of the variance. Females from selected lines lived an average of 19.6 hours longer than females from control lines. There was a significant effect of culture bottle in this experiment. Selection for increased desiccation tolerance has therefore led to a correlated response in starvation tolerance.

To test for an association between desiccation tolerance and ethanol tolerance, lines were characterized for longevity in the presence of 15% ethanol. Flies lived

TABLE 6. Activity and metabolic rate of selected and control lines. Activity was measured in a desiccator (low humidity) and a hydration chamber (high humidity). Means are based on six replicates of 20 flies

		Activity[a]		Metabolic rate
		Low humidity	High humidity	(O_2 mg^{-1} h^{-1})
Mean (S.D.)				
Selected lines	1	11.5 (1.6)	5.5 (1.1)	2.08 (0.15)
	2	10.3 (3.1)	5.8 (2.3)	2.15 (0.31)
	3	9.3 (2.3)	6.8 (1.2)	2.02 (0.32)
Control lines	1	13.2 (2.2)	9.7 (1.0)	2.74 (0.39)
	2	12.5 (1.5)	9.2 (2.3)	2.45 (0.52)
	3	12.2 (1.4)	8.2 (1.2)	2.65 (0.36)
Mean squares for ANOVAs				
Selection (d.f. = 1)		44.4*	78.0**	2.174**
Replicate line (d.f. = 4)		4.3	3.2	0.064
Bottle (d.f. = 6)		1.7	0.4	0.073
Block (d.f. = 2)		10.6	0.3	0.102
Error (d.f.)		4.2 (22)	3.4 (22)	0.147 (16)

[a]The number of flies (out of 15) that moved in a 30-second interval was scored by following five females on three separate occasions (3, 4, and 5 hours after flies were placed in a desiccator); *$P<0.05$; **$P<0.01$.

TABLE 7. Starvation tolerance, ethanol tolerance and fecundity of the selected and control lines. Means and standard deviations are based on six replicates of 20 flies (tolerance) or 23–28 vials (fecundity)

		Starvation (LT_{50}, hours)	Ethanol (LT_{50}, hours)	Fecundity (over 3 days)
Means (S.D.)				
Selected lines	1	105.1 (6.4)	63.6 (12.1)	141.8 (41.5)
	2	94.6 (6.0)	67.6 (9.3)	129.3 (56.4)
	3	94.9 (13.1)	57.4 (11.9)	122.4 (44.0)
Control lines	1	74.8 (4.3)	32.2 (6.7)	158.5 (33.6)
	2	81.5 (5.0)	18.3 (10.4)	162.3 (49.8)
	3	79.5 (6.0)	32.1 (9.5)	166.3 (42.0)
Mean squares for ANOVAs				
Selection (d.f. = 1)		3470.2**	11 252.8**	35 906.2**
Replicate line (d.f. = 4)		143.3	270.4	1614.2
Bottle (d.f. = 6)		154.2***	155.7	—
Error (d.f.)		29.4 (24)	86.7 (24)	2143.8 (119)

**$P<0.01$; ***$P<0.001$.

shorter in this experiment than in the starvation experiment, indicating that the ethanol concentration used was toxic. There was a large significant difference between the selected and control lines (Tabel 7), the variance component due to selection accounting for 83% of the total variance. Selection for increased desiccation tolerance has therefore led to a correlated increase in tolerance to a toxic level of ethanol.

Fecundity

Early fecundity of pairs of males and females was scored over a three-day period. Several flies became stuck in the yeasted medium or escaped during transfer, reducing the number of replicates per line from 30 to 23–28. Means for the three control lines were higher than those for the selected lines (Table 7). On average, females from selected lines laid 31.2 fewer eggs, representing 19% of the eggs laid by females from control lines. An ANOVA indicates a significant effect of selection on fecundity and no effect of replicate line. Selection accounted for 17% of the total variance. Hence early fecundity was reduced by selection.

DISCUSSION

The rapid responses to selection indicate considerable genetic variation for desiccation tolerance in the Melbourne *D. melanogaster* population. This is consistent with earlier studies demonstrating substantial genetic variation in desiccation tolerance (Parsons, 1970) and in rates of water loss (Eckstrand, 1981) among isofemale lines of *D. melanogaster*. The heritability estimates (0.59–0.69) are higher than estimates for life-history traits and most morphological traits in *Drosophila* (cf. Roff & Mousseau, 1987). Populations therefore have the potential to undergo rapid genetic changes when they are exposed to dry habitats.

The increased desiccation tolerance of the selected lines was independent of a number of factors. The absence of changes in body size suggest that changes in the surface-to-volume ratio were not involved. There was no correlated response in the weight loss of anaesthetized flies, suggesting that lines had a similar rate of water loss through the cuticle. The water content of the flies or the water loss flies could tolerate at death were not altered by selection. Water loss via excretion is probably also unimportant because excretory loss forms only a small fraction of the water loss in *D. melanogaster* (Clark & Doane, 1983).

The absence of a correlated response in body size is surprising. Parsons (1970) found a positive correlation between desiccation tolerance and body size of isofemale lines from a *D. melanogaster* population near Melbourne, as did Clark & Doane (1983) in strains differing at the adipose locus. Body size in the selected lines may not have increased because size may only account for a small component of the genetic variance in desiccation tolerance. The absence of a correlated response is consistent with the lack of an association between size and desiccation tolerance at the interspecific level in *Drosophila* (Stanley *et al.*, 1980).

We did find the predicted correlated response in metabolic rate, as evidenced by the lower activity and lower oxygen uptake of the selected lines (Table 6). The decreased water loss of the selected lines after three hours of desiccation suggests that the rate of water loss via the spiracles was altered by selection. It is possible that these correlated responses are related to a common mechanism. For example, a lower metabolic rate may lead to flies keeping their spiracles closed longer because of a reduced demand for gaseous exchange, or because of reduced activity. However, it is not clear from the results whether there is one underlying mechanism or whether several independent mechanisms are involved.

The differences between the selected and control lines for tolerance of ethanol and starvation provide evidence for positive genetic correlations between tolerance of different environmental stresses. Our results are consistent with those of Service *et al.* (1985) in that *D. melanogaster* lines selected for postponed senescence had increased tolerance of desiccation, starvation and a high concentration of ethanol. This suggests that a substantial component of the variance for tolerance of environmental stresses in *Drosophila* may be accounted for by a common tolerance mechanism.

Selected lines had lower fecundity, in agreement with the results of Service *et al.* (1985). Service (1987) found that lines selected for postponed senescence had an increased lipid content, and he suggests that this increase may account for the increased starvation tolerance of these lines as well as their lower early fecundity. He favours this explanation because age-specific changes in lipid content account for the increase in starvation tolerance with age, whereas the metabolic rate of older flies from Service's selected and control lines do not differ. However it is possible that different factors contribute to stress tolerance at different ages. We have examined the lipid content of females from the selected lines in the way described by Service (1987) and found no correlated response for this trait (unpublished data). Thus the decreased fecundity and increased stress tolerance in our selection lines appear more related to metabolic rate than to lipid content.

Negative correlations between life-history traits and stress tolerance may help to explain why tolerance is not higher in the Melbourne population. Desiccation stress is never continuously as extreme in the field as in our laboratory experiments. However geographic variation for this trait follows climatic conditions (Parsons, 1980b) suggesting past directional selection. In addition, the strong correlated selection responses for starvation tolerance indicates that a starvation stress would increase desiccation tolerance. The short lifespan of *Drosophila* in the field (Rosewell & Shorrocks, 1987) suggests exposure to stressful conditions may often occur and there is some evidence that flies are starved under field conditions (Bouletreau, 1978).

The presence of high levels of additive genetic variance for desiccation tolerance also raises the question of how such variance can persist in natural populations. If there are trade-offs between stress tolerance and other fitness components and if insects are regularly exposed to environmental stress, then variation may be maintained when there is overall fitness overdominance (Rose, 1983). Alternatively, *Drosophila* populations may be infrequently exposed to extreme stresses. Levels of genetic variation will then be determined by opposing selection in different generations, high metabolic rate being favoured most of the time, but low metabolic rate being favoured at times of extreme stress. Opposing selection associated with environmental stress has recently been demonstrated for body size in Darwin's finches (Gibbs & Grant, 1987).

In summary, our preliminary results are consistent with our predictions for associations between stress tolerance, metabolic rate and life-history traits. We have found that lines selected for increased desiccation tolerance have decreased activities and decreased metabolic rates. The lines also exhibit increased tolerance to other environmental stresses (ethanol, starvation) and decreased fecundity. We are extending these findings by considering tolerance of other stresses and correlated responses in other life-history traits.

REFERENCES

ATKINSON, D. E., 1977. *Cellular Energy Metabolism and its Regulation.* New York: Academic Press.

BOAG, P. T. & GRANT, P. R., 1981. Intense natural selection in a population of Darwin's finches (Geospinizinae) in the Galapagos. *Science, 214:* 82–84.

BOULETREAU, J., 1978. Ovarian activity and reproductive potential in a natural population of *Drosophila melanogaster. Oecologia, 33:* 319–342.

BUMPUS, H. C., 1899. The variations and mutations of the introduced sparrow, *Passer domesticus. Biology Lectures of the Marine Biology Laboratories, Wood's Hole 1899:* 1–15.

CHARLESWORTH, B., 1984. The evolutionary genetics of life histories. In B. Shorrocks (Ed.), *Evolutionary Ecology:* 117–133. Oxford: Blackwell Scientific Publications.

CLARK, A. G. & DOANE, W. W., 1983. Desiccation tolerance of the adispose[60] mutant of *Drosophila melanogaster. Hereditas, 99:* 165–175.

CLUSTER, P. D., MARINKOVIC, D., ALLARD, R. W. & AYALA, F. J., 1987. Correlations between development rates, enzyme activities, ribosomal DNA spacer-length phenotypes, and adaptation in *Drosophila melanogaster. Proceedings of the National Academy of Sciences, U.S.A., 84:* 610–614.

DANZMANN, R. G., FERGUSON, M. M. & ALLENDORF, F. W., 1988. Heterozygosity and components of fitness in a strain of rainbow trout. *Biological Journal of the Linnean Society, 33:* 285–304.

ECKSTRAND, I. A., 1981. Heritability of water-loss rate in *Drosophila melanogaster. Journal of Heredity, 72:* 434–436.

FAIRBANKS, C. D. & BURCH, G. E., 1970. Rate of water loss and water and fat content of adult *Drosophila melanogaster* of different ages. *Journal of Insect Physiology, 16:* 1429–1436.

FALCONER, D. S., 1981. *Introduction to Quantitative Genetics.* 2nd edn. London: Longman.

GIBBS, H. L. & GRANT, P. R. 1987. Oscillating selection on Darwin's finches. *Nature, 327:* 511–513.

HOCHACHKA, P. W. & SOMERO, G. N., 1984. *Biochemical Adaptation.* Princeton: Princeton University Press.

HOFFMANN, A. A. & COHAN, F. M., 1987. Genetic divergence under uniform selection for knockdown resistance to ethanol in *Drosophila pseudoobscura* populations and their replicate lines. *Heredity, 58:* 425–433.

HORSMAN, D. C., ROBERTS, T. M. & BRADSHAW, A. D., 1979. Studies on the effect of sulphur dioxide on perennial ryegrass *(Lolium perenne* L.). *Journal of Experimental Botany, 30:* 495–501.

HUNTER, A. S., 1964. Effects of temperature on *Drosophila* I. Respiration of *D. melanogaster* grown at different temperatures. *Comparative Biochemistry & Physiology, 11:* 411–417.

IVANOVICI, A. M. & WIEBE, W. J., 1981. Towards a working 'definition' of stress: a review and critique. In G. W. Barrett & R. Rosenberg (Eds), *Stress Effects on Natural Ecosystems:* 13–27. New York: John Wiley.

KOEHN, R. K. & SHUMWAY, S. R., 1982. A genetic/physiological explanation for differential growth rate among individuals of the American oyster *Crassotrea virginica* (Gmelin). *Marine Biology Letters, 3:* 35–42.

LANDE, R. & ARNOLD, S. J., 1983. The measurement of selection on correlated characters. *Evolution 37:* 1210–1226.

LANGRIDGE, J. & GRIFFING, B., 1959. A study of high temperature lesions in *Arabidopis thaliana. Australian Journal of Biological Sciences, 12:* 117–135.

LAURIE-AHLBERG, C. C., BARNES, P. T., CURTSINGER, J. W., EMIGH, T. H., KARLIN, B., MORRIS, R., NORMAN, R. A. & WILTON, A. N., 1985. Genetic variability of flight metabolism in *Drosophila melanogaster* II. Relationship between power output and enzyme activity levels. *Genetics, 111:* 845–868.

MACHADO-ALLISON, C. E. & CRAIG, G. B. Jr., 1972. Geographic variation in resistance to desiccation in *Aedes aegypti* and *A. atropalpus* (Diptera: Culicidae). *Annals of the Entomology Society of America, 65:* 542–547.

MARINKOVIC, D., MILOSEVIC, M. & MILANOVIC, M., 1986. Enzyme activity and dynamics of *Drosophila* development. *Genetica, 70:* 43–52.

MATHESON, A. C. & PARSONS, P. A., 1973. The genetics of resistance to long-term exposure to CO_2 in *Drosophila melanogaster*, an environmental stress leading to anoxia. *Theoretical & Applied Genetics, 42:* 261–268.

MITTON, J. B., CAREY, C. & KOCHER, T. D., 1986. The relation of enzyme heterozygosity to standard and active oxygen consumption and body size of tiger salamanders, *Ambystoma tigrinum. Physiological Zoology, 59:* 574–582.

MIYOSHI, Y., 1961. On the resistability of *Drosophila* to sodium chloride I. Strain differences and heritability in *D. melanogaster. Genetics, 46:* 935–945.

MUIR, W. M., 1986. Estimation of response to selection and utilization of natural populations for additional information and accuracy. *Biometrics, 42:* 381–391.

NEVO, E., BEILES, A. & BEN-SHLOMO, R., 1984. The evolutionary significance of genetic diversity: ecological, demographic and life-history correlates. In G. S. Mani (Ed.), *Evolutionary Dynamics of Genetic Diversity:* 13–21, Berlin: Springer Verlag.

NEVO, E. & SHKOLNIK, A., 1974. Adaptive metabolic variation of chromosome forms in mole rats *Spalax. Experientia, 30:* 724–726.

NIX, H. A., 1981. The environment of Terra Australis. In A. Keast (Ed.), *Ecological Biogeography of Australia:* 103–113. The Hague: Junk.

OGAKI, M. & NAKASHIMA-TANAKA, E., 1966. Inheritance of radioresistance in *Drosophila. Mutation Research, 3:* 438–443.

PARSONS, P. A., 1969. A correlation between the ability to withstand high temperature and radioresistance in *Drosophila melanogaster. Experientia, 25:* 1000.

PARSONS, P. A., 1970. Genetic heterogeneity in natural populations of *Drosophila melanogaster* for ability to withstand desiccation. *Theoretical & Applied Genetics, 40:* 261–266.

PARSONS, P. A., 1974. Genetics of resistance to environmental stresses in *Drosophila* populations. *Annual Review of Genetics, 7:* 239–265.

PARSONS, P. A., 1980a. Isofemale strains and evolutionary strategies in natural populations. *Evolutionary Biology, 13:* 175–217.

PARSONS, P. A., 1980b. Adaptive strategies in natural populations of *Drosophila:* ethanol tolerance, desiccation resistance, and development times in climatically optimal and extreme environments. *Theoretical & Applied Genetics, 57:* 257–266.

PARSONS, P. A., 1981. Evolutionary ecology of Australian *Drosophila:* A species analysis. *Evolutionary Biology, 14:* 297–350.

PARSONS, P. A., 1983. *The Evolutionary Biology of Colonizing Species.* New York: Cambridge University Press.

PARSONS, P. A., 1987a. Evolutionary rates under environmental stress. *Evolutionary Biology, 21:* 311–357.

PARSONS, P. A., 1987b. Features of colonizing animals: phenotypes and genotypes. In A. J. Gray, M. J. Crawley & P. J. Edwards (Eds), *Colonization, Succession and Stability:* 133–154. Oxford: Blackwell Scientific Publications.

PARSONS, P. A., MACBEAN, I. T. & LEE, B. T. O., 1969. Polymorphism in natural populations for genes controlling radioresistance in *Drosophila. Genetics, 61:* 211–218.

PLAPP, F. W., 1976. Biochemical genetics of insecticide resistance. *Annual Review of Entomology, 21:* 179–197.

ROFF, D. A. & MOUSSEAU, T. A., 1987. Quantitative genetics and fitness: lessons from *Drosophila. Heredity, 58:* 103–118.

ROSE, M. R., 1983. Theories of life-history evolution. *American Zoologist, 23:* 15–23.

ROSE, M. R. & CHARLESWORTH, B., 1981. Genetics of life history in *Drosophila melanogaster* II. Exploratory selection experiments. *Genetics, 97:* 187–196.

ROSEWELL, J. & SHORROCKS, B., 1987. The implication of survival rates in natural populations of *Drosophila:* capture-recapture experiments on domestic species. *Biological Journal of the Linnean Society, 32:* 373–384.

SERVICE, P. M., 1987. Physiological mechanisms of increased stress resistance in *Drosophila melanogaster* selected for postponed senescence. *Physiological Zoology, 60:* 321–326.

SERVICE, P. M., HUTCHINSON, E. W., MACKINLEY, M. D. & ROSE, M. R., 1985. Resistance to environmental stress in *Drosophila melanogaster* selected for postponed senescence. *Physiological Zoology, 58:* 380–389.

SERVICE, P. M. & ROSE, M. R., 1985. Genetic covariation among life-history components: the effects of novel environments. *Evolution, 39:* 943–945.

SOKAL, R. R. & ROHLF, F. J., 1981. *Biometry.* 2nd ed. New York: Freeman.

SOUTHWOOD, T. R. E., 1988. Tactics, strategies and templets. *Oikos 52:* 3–18.

STANLEY, S. M., PARSONS, P. A., SPENCE, G. E. & WEBER, L., 1980. Resistance of species of the *Drosophila melanogaster* subgroup to environmental extremes. *Australian Journal of Zoology, 28:* 413–421.

STARMER, W. T., HEED, W. B. & ROCKWOOD-SLUSS, E. J., 1977. Extension of longevity in *Drosophila mojavensis* by environmental ethanol: Differences between subraces. *Proceedings of the National Academy of Sciences, U.S.A., 74:* 387–391.

TAYLOR, G. E., 1978. Genetic analysis of ecotypic differentiation within an annual plant species *Geranium carolinianum* L. in response to sulphur dioxide. *Botanic Gazette, 139:* 362–368.

TURNER, R. G., 1969. Heavy-metal tolerance in plants. In I. H. Rorison (Ed.), *Ecological Aspects of the Mineral Nutrition of Plants:* 399–410. Oxford: Blackwell Scientific Publications.

WATT, W. B., 1985. Bioenergetics and evolutionary genetics: opportunities for new synthesis. *American Naturalist, 125:* 118–143.

WADDINGTON, C. H., 1959. Canalization of development and genetic assimilation of acquired characters. *Nature, 183:* 1654–1655.

WEBB, L. J., TRACEY, J. G. & WILLIAMS, W. T., 1984. A floristic framework of Australian rainforests. *Australian Journal of Ecology, 9:* 169–198.

WESTERMAN, J. M. & PARSONS, P. A., 1973. Variation in genetic architecture at different doses of Co^{60}-radiation as measured by longevity in *Drosophila melanogaster. Canadian Journal of Genetics & Cytology, 15:* 289–298.

WHITE, E. B., DEBACH, P. & GARBER, M. J., 1970. Artificial selection for genetic adaptation to temperature extremes in *Aphytis lingnanensis* Compere (Hymenoptera: Aphelinidae). *Hilgardia, 40:* 161–192.

Biological Journal of the Linnean Society (1989), *37*: 137–155. With 8 figures

Evolution and stress—genotypic and phenotypic components

A. D. BRADSHAW AND K. HARDWICK

Department of Environmental and Evolutionary Biology, University of Liverpool, Liverpool L69 3BX

Since stress can be defined as anything which reduces growth or performance, it follows that, if appropriate genetic variability is present, classical evolutionary changes in populations are to be expected in any situation where a consistent stress is occurring. There is now considerable evidence for such evolution, producing constitutive adaptations in plants in response to stress, which are specific to the stress concerned. Stress may however operate in a temporary or fluctuating manner. In these situations, facultative adaptations, able to be produced within a single genotype through phenotypic plasticity, will be more appropriate. Very different specific phenotypic response systems, both morphological or physiological, can be found in plants in relation to different fluctuating stresses, operating over a wide range of time scales. These response systems are under normal genetic control and appear to be products of normal evolutionary processes. They can however have quite complex features, analogous to the behavioural response systems in animals.

KEY WORDS:—Evolution and stress – constitutive adaptations – facultative adaptations – phenotypic plasticity – morphological response systems – physiological response systems.

CONTENTS

INTRODUCTION

It was our original intention, in discussing the processes of evolution in response to stress, to consider both plants and animals. In the event we have found that what can be discussed in relation to plants alone is so extensive that, to give what we hope is a critical picture of what can occur, we have decided to restrict our palette to plants. This may disappoint some of our zoological colleagues, although it might be argued that there are already too many treatments of evolution that are zoological. Perhaps this contribution may

0024–4066/89/050137 + 19 $03.00/0

redress the balance a little, as well as emphasizing the wealth of distinctive evidence which plants can offer to an understanding of the significance of stress in evolution.

THE CLASSICAL PICTURE—CONSTITUTIVE ADAPTATION

If we assume that stress is anything which reduces growth or performance, it follows that, in a situation where a particular stress operates, there must be a reduction in fitness—defined in the normal Darwinian sense as the ability to contribute to the next generation. If genotypes in a population vary in resistance or tolerance to this stress, then the differences in relative fitness which will accrue must lead to genetic change in the population in the direction of increased resistance or tolerance to the particular stress. In other words, *normal, classical, evolutionary changes in populations are to be expected in any situation where stress is occurring consistently;* indeed they are inevitable, providing that appropriate genetic variability exists in the populations affected. This latter qualification is most important and we shall return to it.

There are two aspects of this classical, genotypic, evolution that can be profitably studied, first the mechanism by which the evolution occurs, and secondly the results it produces. The first will give us an idea of what can happen; the second will tell us what actually does happen.

Evolutionary mechanisms

There are plenty of theoretical treatments of evolution, but what is important is to see what happens in practice. Perhaps one of the situations in which we can best see evolution at work as a result of stress is when there has been contamination of soils by heavy metals such as copper, zinc or lead. The situation has been well studied and reported in detail and in major reviews (Antonovics, Bradshaw & Turner, 1971; Macnair, 1981; Baker, 1987) so that only an outline of the salient features is necessary here. Metal contamination has two advantages for the study of evolutionary mechanisms, firstly that as a stress factor it is usually relatively stable and long lasting, and secondly that it is usually relatively uncomplicated by other factors, although these may sometimes occur in parallel.

Metal contamination, from whatever source, can cause very severe damage to vegetation and even lead to its complete death, because there appear to be very few species which have a pre-existing adaptation to tolerate heavy metal toxicity. However, in most sites where metal contamination has been present for a few years, one or two species can usually be found growing successfully. It turns out that this is because they possess, and must have evolved, metal-tolerant populations. This tolerance is on the whole specific to individual metals, and is highly heritable when tested by parent/offspring regressions and diallel crosses (e.g. Gartside & McNeilly, 1974).

The origin of this tolerance appears to be from a very low frequency (about 2 per 1000) of tolerant or partially tolerant individuals which are to be found in the base populations of these species (Walley, Khan & Bradshaw, 1974). In toxic conditions, only these tolerant individuals survive. There is then further selection

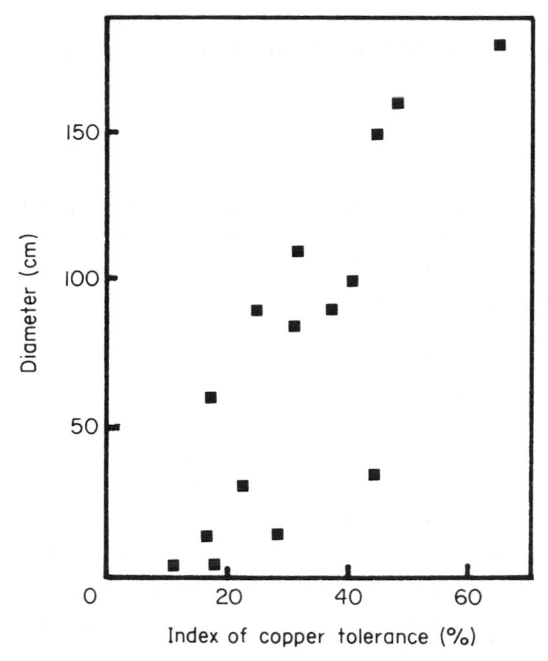

Figure 1. The relationship between growth (diameter) and copper tolerance of fifteen plants of *Agrostis stolonifera* surviving on copper-contaminated soil at a copper refinery at Prescot, Merseyside after all other plants had been eliminated by toxicity (from Wu, Bradshaw and Thurman, 1975).

among the survivors; the most tolerant individuals grow more successfully, and therefore contribute most to the next generation (Fig. 1). Since the tolerance is heritable, the offspring form the nucleus of a newly evolved tolerant population. The initial tolerance may not be complete and the paucity of the initial survivors may mean that two or three generations must elapse before a fully tolerant population is properly established. However, this does not prevent markedly tolerant populations, able to cope with substantial toxicity, establishing in only a few years. A good example of rapid evolution of tolerance is the copper-tolerant populations established in the neighbourhood of the copper refinery at Prescot, Merseyside (Wu, Bradshaw & Thurman, 1975). Similar speeds of evolution, in response to the stress of zinc toxicity, are found beneath electricity pylons (Al-Hiyaly, McNeilly & Bradshaw, 1988) and other recent sources of zinc contamination (Ernst, 1976).

There is, however, one important rider: such evolution only takes place in certain species. Other species, despite being subjected to the same level of stress, do not evolve metal tolerance and are eliminated from these toxic sites. Careful analysis shows that this is because they do not possess the appropriate genetic variation in their normal populations (Bradshaw, 1984). Such a limitation, set by the genetic constitution of the starting material, is particularly easy to recognize for a character as distinctive as metal tolerance, but the limitation has been well known for many decades by plant breeders, who may have to take complicated steps to overcome it.

Adapted products

Metal tolerance illustrates the mechanism of evolution of constitutive adaptation to one stress. But what range of adaptations to other stresses can be achieved? To provide convincing evidence, the material must be the product of evolution in relation to recent or existing stresses, where it is unlikely that the products are due to some historic accident or other factor unconnected with the stress operating.

Such evidence is best provided by populations from within a single species, although pairs of closely related species can also be informative (Bradshaw, 1987). There is in fact a wealth of examples of population differences. These have mostly been described by ecological geneticists rather than physiologists. Nevertheless it is very clear that very distinctive characteristics can evolve, each related to the nature of the particular stress operating (Table 1). There is little sign of any generalized evolutionary response to stress (Grime, 1977).

There are, however, few investigations of the availability of the appropriate genetic variation within the species subjected to the stresses in the examples given in Table 1. For herbicide resistance there is evidence of restrictions to evolution due to lack of variability (LeBaron & Gressel, 1982). For salt tolerance, however, it appears that variation exists in all of a wide range of species examined (Ashraf, McNeilly & Bradshaw, 1986a). We must not forget that when a stress is applied to a population, variation previously unselected can be uncovered, called "hidden variation" by Cooper (1954) who first drew attention to it.

As with the desiccation resistance in *Drosophila* described by Hoffmann & Parsons (see p. 117), what is crucial before evolution can occur in response to a stress is that the variability should be heritable. Persistence into offspring is in itself good evidence, but precise determination of heritability requires the use of parent/progeny regressions or diallel analyses (Lawrence, 1984). These show very high heritabilities for metal tolerance—well over 50%. Salt tolerance can show similar levels of heritability (Table 2), although narrow sense heritability measuring additive variation is lower than the broad sense heritability measuring total genetic variation (Ashraf, McNeilly & Bradshaw, 1986b).

TABLE 1. Examples of different types of stress and the different constitutive adaptations to them

Stress	Adaptation	Example
Metal contamination	Chemical complexing	*Agrostis, Silene, Mimulus*[a]
Low nutrients	More efficient uptake	*Festuca ovina*[b]
Wind	Dwarfing etc.	*Agrostis stolonifera*[c]
Cold	Freezing control	Many species[d]
High light	Elevated RUBISCO etc.	*Solidago virgaurea*[e]
Salt	Osmoregulation etc.	*Enteromorpha intestinalis*[f]
Triazine herbicides	Chloroplast membrane change	Many species[g]

[a]Woolhouse, 1983; [b]Snaydon & Bradshaw, 1961; [c]Aston & Bradshaw, 1966; [d]Larcher & Bauer, 1981; [e]Bjorkman & Holmgren, 1963; [f]Young, Collins & Russell, 1987; [g]Arntzen, Pfister & Steinback, 1982.

TABLE 2. Estimates of broad and narrow sense heritability of salt tolerance in different populations of various grass species (Ashraf, McNeilly & Bradshaw, 1986b)

Species	Site	NaCl test concentration (mol m^{-3})	Narrow sense heritability (%)	Broad sense heritability (%)
Agrostis stolonifera	Abraham's Bosom	200	32	94
Holcus lanatus	Abraham's Bosom	150	41	88
Dactylis glomerata	Abraham's Bosom	150	54	93
Dactylis glomerata	Foryd Bay	150	56	90
Holcus lanatus	Aberdesach	150	29	96
Festuca rubra	Parkgate	150	29	84

So evolution of constitutive adaptations to stress would not seem to be difficult, and we should expect very rapid evolutionary changes under appropriate conditions so long as the variability is available. There are however limits which not only determine whether any evolution can actually occur, but also affect how far the evolution can proceed. Thus in metal-contaminated sites, areas can always be found where there are no plants growing, because the metal levels are higher than any metal tolerance system so far evolved can cope with. The same is true in saline areas. Nevertheless the evolution which does occur, considerably increases the ecological amplitude of the species concerned, since without its occurrence the populations of the species would be limited to a much narrower range of habitats.

At the same time, it must be realized that what evolution takes place depends on the details of the stress situation, This is well illustrated by the salt tolerance exhibited by different grass populations (Ashraf *et al.*, 1986b). In this material suprisingly, the sea cliff population of *Agrostis stolonifera* at Abraham's Bosom does not show clear tolerance to salt in solution surrounding its roots, although other populations do and the Abraham's Bosom population does contain heritable variation for the character. Instead the population shows resistance to salt spray, a capacity which appears to be due to its very distinctive leaf epidermis (Fig. 2) (McNeilly, Ashraf & Veltkamp, 1987). All this relates to the fact that although the cliff population is very exposed to salt spray, it occupies a soil kept effectively salt-free by freshwater drainage. Adaptations to stress can be very specific.

This raises a major problem. It is clear that stress can have many different manifestations which require different adaptations. First, many quite different environmental factors can cause stress. Secondly, any one stress factor can vary in time or space; we have not considered this, for we have chosen situations where the stress is effectively permanent and unchanging. Yet it is not at all unusual for an environmental factor, for instance light, to vary markedly from one time to another or from one place to another, so that stress occurs at some times or in some places and not others.

The result of this is that a single individual may experience quite different environments on different occasions. Equally, parents and their offspring may experience different environments. In such situations normal evolution by constitutive changes cannot provide the optimal adaptive solution. What is

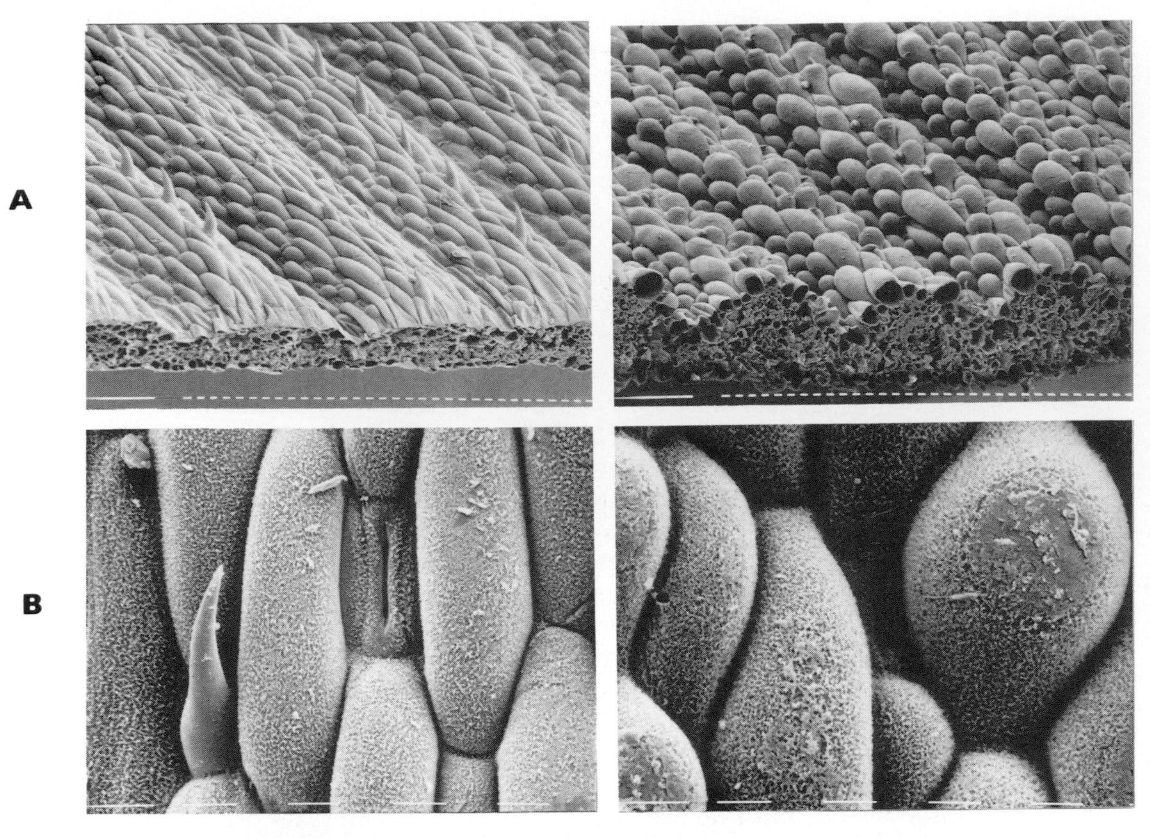

Figure 2. A, Transverse sections and views of the adaxial surface; B, close-ups of stomata on the adaxial surface, of leaves of *Agrostis stolonifera* taken from a normal pasture (left) and a sea cliff at Abraham's Bosom, Anglesey (right), using scanning electron microscopy (from McNeilly, Ashraf and Veltkamp, 1987).

required is the potential to show *adaptive phenotypic responses within the individual genotype,* able to occur at a speed which matches the speed of environmental change. These will have to occur without any parallel genetic change, since this is ruled out by the stability of the genetic system in somatic tissue.

A plasticity of the phenotype rather than of the genotype is required. This can indeed occur. It has its own distinct set of characteristics, quite different from those applying to constitutive adaptations, which require their own detailed study.

THE ALTERNATIVE PICTURE—FACULTATIVE ADAPTATION

The idea that phenotypic plasticity could be an evolutionary pathway in certain circumstances is not new. It was foreshadowed by Darwin in a letter to Karl Semper in 1881: "I speculated whether a species very liable to repeated and great changes of conditions might not assume a fluctuating condition ready to be adapted to either condition".

TABLE 3. Characteristics of phenotypic plasticity

(a) Specific for individual characters
(b) Specific in relation to particular environmental influences
(c) Specific in pattern and direction
(d) Under additive genetic control

Since that time a great deal of evidence about the characteristics of phenotypic plasticity and the situations in which it is likely to be manifest has been gathered together. There are three pertinent reviews (Bradshaw, 1965; Schlichting, 1986; Sultan, 1987). The essential features are summarized in Table 3. It is important to realize that although the adaptive *response* is facultative and can appear 'on demand', the *response system* is constitutive and is evolved by the normal, classical, evolutionary processes we have aleady considered.

On the whole, the main interest until recently has been in morphological rather than physiological plasticity. This must be partly because some of the examples of morphological plasticity, such as the heterophylly occurring in aquatic plants, e.g. in *Ranunculus peltatus* and *Sagittaria sagittifolia,* are both spectacular and remarkable. But it must also be because, until recently, plant physiologists have been more interested in average performance rather than the capacity of plants to respond to variations in their environment which might thereby optimize their performance. The situation is very different now, as witnessed by recent publications (e.g. Smith, 1981; Jennings & Trewavas, 1986).

Adaptive response systems

Although the ways in which adaptation is achieved in constitutive and facultative systems are fundamentally different, there are not necessarily any differences in the actual physiological mechanisms involved. Apparently constitutive mechanisms can have facultative-response elements. This is very clear in the salt tolerance shown by different populations of *Enteromorpha intestinalis* which grows in both marine and estuarine situations (Young, Collins & Russell, 1987). The marine populations have thick cell walls and high levels of organic solutes, particularly dimethyl sulphoniopropionate (DMSP); in contrast, the estuarine populations have thin cell walls and low levels of DMSP. However, the latter have an ability, not possessed by the marine populations, to alter DMSP levels in relation to external salinity, and can thereby adapt to the fluctuating salinity levels to which they are subject in estuarine situations (Fig. 3). Another example is seen in the woodland and tundra populations of *Solidago virgaurea* (Bjorkman & Holmgren, 1963). Their mean photosynthetic responses to light intensity are very different. The tundra populations, however, which experience great annual variation in their light climates not experienced by woodland populations, also have the ability to alter their response depending on the light conditions to which they have recently been subjected. This physiological plasticity is completely absent in the woodland populations. Such differences are discussed further by Mooney & Gulmon (1979).

There is a quite remarkable range of phenotypic responses to stress, from changes at the molecular level to changes in the growth of the whole plant.

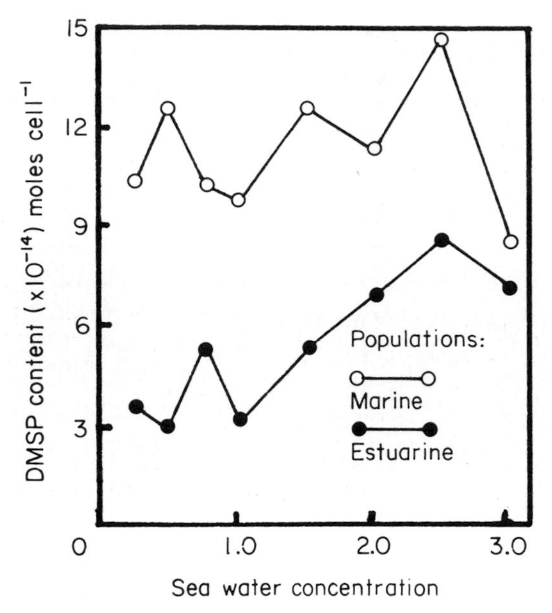

Figure 3. The response of intracellular levels of DMSP, expressed per cell, to different levels of salinity, in populations of *Enteromorpha intestinalis* from different environments (from Young, Collins and Russell, 1987).

Table 4 shows some examples in relation to variation in one factor, light, but there are many others.

It is always possible that the response systems are the result of some evolutionary accident and our thesis of attributing adaptive significance to them, fallacious—we could be falling into the '*a posteriori*' trap. However they do have remarkable ecological logic. They also occur only in some species or populations and not in others, although these are closely related. This is good evolutionary evidence that the response systems have been produced by normal evolutionary processes in relation to the environments in which the species or populations occur. This is very clear for the phytochrome control of stem extension, where woodland species in general show less response than species from open habitats, and the species which persists in the most dense shade, *Mercurialis perennis,* shows almost no response at all (Table 5) (Morgan, 1981).

TABLE 4. Different types of phenotypic response to light

Nature of light variation	Mechanism	Response type*	Time scale
Sunflecks	Grana stacking[a]	R	Seconds-minutes
Diurnal	Chloroplast movement[b]	R	Minutes-hours
Change of direction	Leaf orientation[c]	R or I	Minutes-days
Spatial	Etiolation[d]	I	Hours-days
Seasonal-spatial	Leaf development[e] (sun/shade)	I	Days

*Reversible (R), irreversible (I); [a]Anderson, 1986; [b]Virgin, 1964; [c]Lang & Begg, 1979; [d]Morgan, 1981; [e]Lichtenthaler, 1985.

TABLE 5. Contrasting stem extension of different species under simulated woodland conditions (+supplementary FR) (derived from Morgan, 1981)

Species	Relative internode extension response	Collection site
Senecio vulgaris	100	Arable land
Chenopodium album	67	Arable land
Sinapis alba	62	Arable land
Circaea lutetiana	54	Woodland
Urtica dioica	37	Hedgerow
Teucrium scorodonia	19	Woodland
Mercurialis perennis	5	Woodland

A further line of evidence that suggests that the response systems have such ecological, adaptational, logic to confirm that they must be the products of normal evolutionary processes, is manifest if we examine how plants can adapt to environmental changes occurring in time, but with quite different time scales. The different mechanisms of response to light, given in Table 4 for instance, have very different speeds of response, each appropriate to the different types of light variation which can occur. By this, plants can optimize very precisely the use of light available to them.

As already noted, many different factors are capable of causing stress, and this is as relevant for facultative adaptation as for constitutive adaptation. It is not possible to describe the full range of mechanisms known, but Table 6 gives examples, for different stress factors, of populations or closely related species showing marked differences in phenotypic response.

In some cases some populations have an ability to alter their behaviour in response to stress and others do not. Sand dune populations, but not fertile grassland populations, of *Capsella bursa-pastoris* can advance their flowering time under drought and nutrient stress (Sørensen, 1954). Populations of *Ranunculus flammula* from temporary lakes can alter their leaf shape on exposure to air, but not populations from permanent lakes (Cook & Johnson, 1968). Perhaps one of

TABLE 6. Examples of different types of fluctuating stress and different facultative adaptations to them

Factor	Mechanism	Response type	Time scale	Example
Water	Stomata	S/R	Minutes	*Sempervivum*[a]
Predation	Phytoalexins	S/R	Hours	Many species[b]
Temperature	Leaf growth	S/I	Days	*Lolium perenne*[c]
Nutrients	Root growth	S/I	Days	*Agrostis stolonifera*[d]
Drought	Leaf abscission	S/I	Days	*Theobroma cacao*[e]
Vegetation height	Petiole length	S/I	Days	*Trifolium repens*[f]
Season	Leaf abscission	C/I	Weeks	*Quercus* sp.[g]
Submergence	Leaf development	C/I	Weeks	*Ranunculus flammula*[h]
General stress	Flowering time	C/I	Weeks	*Capsella bursa-pastoris*[i]

Simple (S), complex (C); reversible (R), irreversible (I); [a]Losch & Tenhunen, 1981; [b]Darvill & Albersheim, 1984; [c]Cooper, 1964; [d]Grime, Crick & Rincon, 1986; [e]Alvim *et al.*, 1974; [f]Kerner von Marilaun & Oliver, 1895; [g] Tucker, 1952; [h] Cook & Johnson, 1968; [i]Sørensen, 1954.

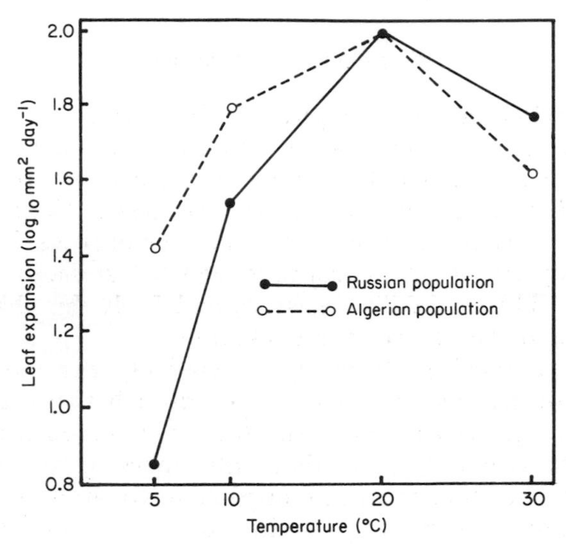

Figure 4. The contrasting responses in leaf expansion of two populations of ryegrass (*Lolium perenne*) to different temperatures (from Cooper, 1964).

the first examples noted is the remarkable ability of *Trifolium repens* to elongate its petiole referred to by Kerner von Marilaun & Oliver (1895). Many other members of the genus do not have the same ability.

In other cases all populations may show response to stress, but the nature of the responses are different. A remarkable example is leaf area development in response to temperature in the grasses *Lolium perenne* and *Dactylis glomerata* (Cooper, 1964). All populations of both species show the normal relationship of leaf growth to temperature with a maximum at 20°C. However, Mediterranean populations show relatively good growth at low temperatures and poor growth at high temperature whereas N. European populations show very poor growth at low temperature and relatively good growth at high temperature (Fig. 4). At first sight these responses might appear to be 'the wrong way round', until it is realized that in the Mediterranean the cool winter period is the season when leaf growth is advantageous because it is not a season of temperature or moisture stress, in contrast to the summer, whereas in N. Europe cool temperatures are a signal of an extremely high stress winter period to follow.

Some response systems are very complex. Nowhere is this more apparent than in phytoalexin production. Many forms of plant defence, both morphological and chemical, against pathogens and insect attack, are constitutive and always present in those species that possess them. This is, of course, a wasteful system since metabolic energy has to be used to produce the character even when it is not required (see Sibly & Calow, p. 101). Phytoalexins are, in contrast, only produced when pathogen attack occurs and are a remarkable, if rather specialized, type of phenotypic plasticity. There are considerable differences between different species and cultivars in their ability to produce phytoalexins, and the production is usually itself specific for particular races of the pathogen. Phytoalexin production therefore depends on interaction between the genotype of both host and pathogen, a very particular form of stress response (Bailey, 1987).

Evolutionary mechanisms

There are therefore many examples of phenotypic response systems that provide facultative adaptation to transient stresses. From the occurrence of these systems in some species or populations and not others we can argue that they must be the product of normal evolutionary processes. Certainly the selection required is present—as an outcome of the existence of stress. What is not so self evident is the occurrence of appropriate genetic variability on which this selection can act. This variability must control both the ability of plants to respond to stress and the nature of this response.

Herein lies a practical problem. It is relatively easy to observe genetic variation in constitutive characters by comparison between single individuals grown under a single environment (in this case an environment with the appropriate stress operating), so that differences between individuals in tolerance to a particular stress will be immediately apparent. However, when we are looking for genetic variability in the ability to respond in an adaptive manner to stress, it is necessary to compare the behaviour of the single individuals under two environments, the normal and the stressed. This is more difficult and can only be achieved through clonal propagation of the individuals, or the use of different samples of different pure lines.

As a result, there is much less evidence for genetic variability in response. However, cultivars of inbreeding or vegetatively propagated crops provide ideal material since these are effectively single genotypes and can easily be grown and compared in different environments. It can, for instance, be easily shown that many cultivars of lettuce other than Grand Rapids do not show the phytochrome mediated response to light quality which triggers seed germination.

It has been appreciated for a long time that different cultivars of crop plant differ in their tolerance to environmental stresses, e.g. in tomato, to different growing conditions (Williams, 1960), and similarly in tobacco (Jinks & Mather, 1955). Sometimes, but not necessarily, this may be related to heterozygosity (Allard & Bradshaw, 1964). What is interesting is to look at the detailed response of a number of cultivars to the stress caused by variation in a single environmental factor. In sunflower considerable fixed differences can be found between cultivars, especially between wild and cultivated, but only in some characters, such as seed size. At the same time, however, in response to density stress one character can be held stable in all cultivars (leaf number) while another varies (leaf size); and another varies in some cultivars but not others (seed size) (Bradshaw, 1973) (Fig. 5). The detailed pattern of response of a single character (height) can also be significantly different between different cultivars (Fig. 6). It would be interesting to speculate on the mechanisms involved in such response differences. The crucial point, however, is that genetic variability in response to stress does exist.

To define this variability more closely requires breeding experiments by which the additive, heritable, component can be determined. On the few occasions when this has been done, e.g. by parent/offspring regression for flowering time variation in different environments in *Bromus mollis* (Jain, 1978), by diallel analysis of character variation in different environments in *Arabidopsis thaliana* (Perkins & Jinks, 1973), or by analysis of F_3 families for response to density in

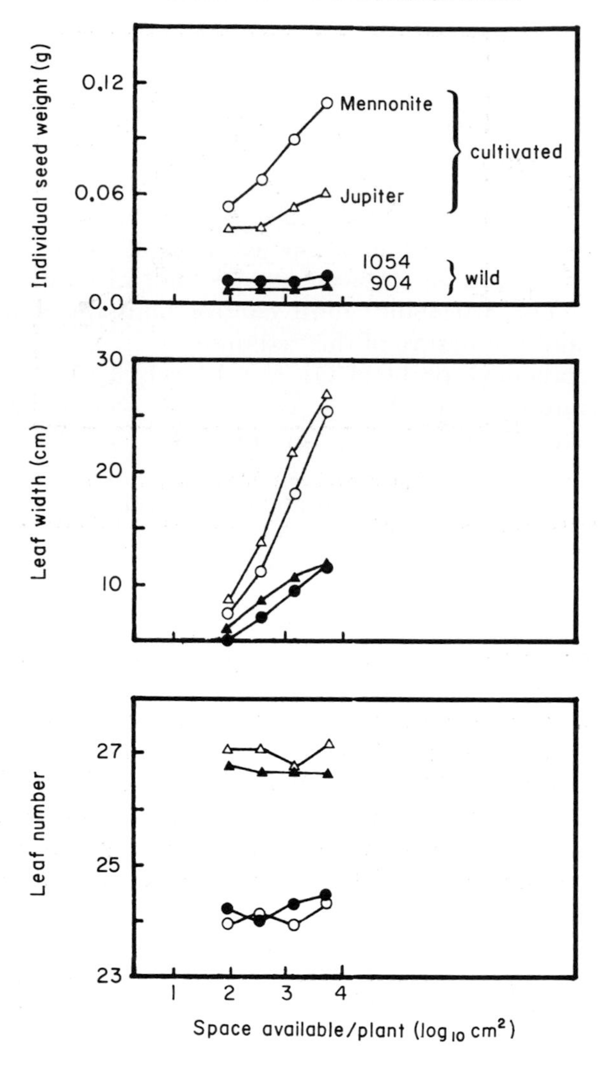

Figure 5. The contrasting responses of various characters in different varieties of sunflower (*Helianthus annus*), both wild and cultivated, to variation in density (from Khan, 1967).

TABLE 7. Heritability of response to density of individual characters in *Linum usitatissimum* in F_3 families derived from crosses between cultivars (Khan, Antonovics & Bradshaw, 1976)

	Redwing × Weira	Redwing × Maroc
Height	0.45***	0.76***
Capsules	0.63***	0.73***
Branches	0.51***	0.68***
Dry weight	0.60***	0.68***

***$P<0.001$.

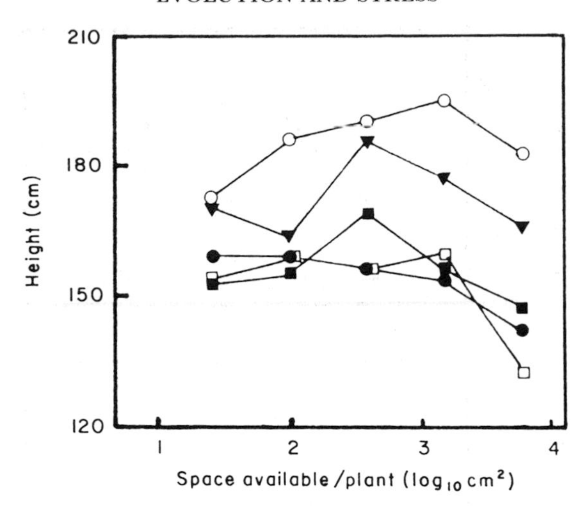

Figure 6. The different responses in height of five cultivated varieties of sunflower to variation in density (from Khan, 1967).

Linum usitatissimum (Khan, Antonovics & Bradshaw, 1976), (Table 7), high heritabilities have been found.

What is critical to the evolution of complex responses is that the response of one character can show independence from that of another. Although this can differ in different species (Schlichting, 1986), it suggests that pleiotropy does not have to underly the response of different characters. This is emphasized further in crossing experiments which show that there is segregation of the responses of different characters (Khan *et al.*, 1976). In crosses between different cultivars of

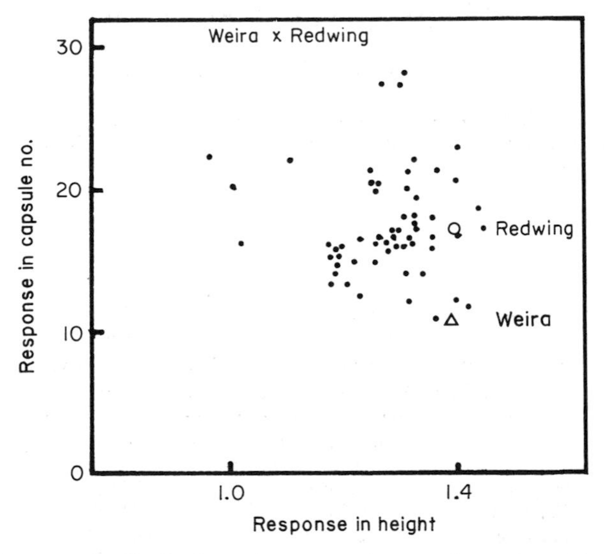

Figure 7. The independent distributions of responses (measured as value at low/value at high density) for capsule number and height of a random set of sixty F_3 lines from two crosses between different cultivars of *Linum usitatissimum* (from Khan, Antonovics & Bradshaw, 1976).

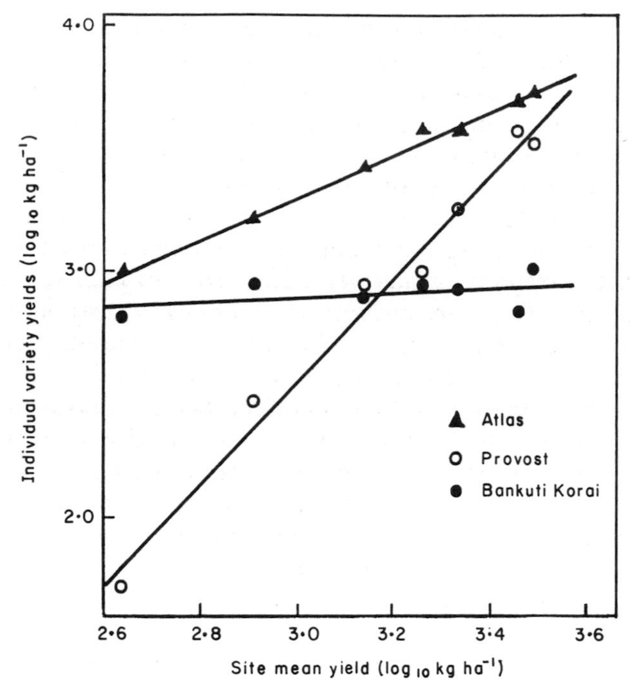

Figure 8. The contrasting responses in yield of three cultivars of barley to a range of different environments, the impact of which has been assesed by the mean yield of a large number of cultivars tested (from Finlay & Wilkinson, 1963).

Linum usitatissimum the responses in capsule number and height show little correlation in the F_3 generation (Fig. 7). So new combinations of responses can be produced by selection.

It is disappointing that, although in animals, particularly in *Drosophila,* selection for (or against) response to stress has been carried out (particularly the early work of Waddington (1960) and see Hoffman & Parsons, p. 117), there does not appear to be any fully recorded work on plants. However plant breeders have taken selection for stability of yield, and sometimes selection for other characters such as lack of premature flowering in sugar beet, as a crucial aspect of their work. This work has been helped by the approach advocated by Finlay & Wilkinson (1963) in which the response of any single cultivar to stress is indicated by the regression over many different environments of the yield of that cultivar against the yield of a series of cultivars grown with it in the same series of environments. This can reveal considerable differences in tolerance to stress (Fig. 8). This approach is now widely used to assess material in plant breeding programmes (e.g. Matsuo, 1975), although other methods of assessing genotypic responses to environments have also been developed (Westcott, 1986).

Some physiological considerations

Physiologists have revealed very considerable subtleties and complexities in the facultative responses of plants to stress. Indeed, although quite different mechanisms are involved, there are close analogies with what occurs in animals, and it is appropriate to refer to different sorts of plant 'behaviour', for instance foraging behaviour in *Glechoma hederacea* (Slade & Hutchings, 1987). The species is able to elongate its stolons and limit branching in areas where light intensity and nutrient concentrations are low, and thereby enhance the likelihood of finding more favourable areas. The ways in which plants forage for nutrients has recently begun to attract more attention (Grime, Crick & Rincon, 1986).

All behavioural systems involve a stimulus and a response. Normally in animals there is a close adaptive connection between these; the stimulus is the factor to which the response is adaptive. However, in plants, this connection is not always so simple. There are indirect as well as direct response systems. Direct systems of response are familiar, e.g. the closure of stomata with drought, the increased growth of roots where there are high concentrations of nutrients, or the decreased growth of plants at high density. Indirect systems, however, are equally commonplace. The shedding of leaves of deciduous trees is not brought about by the cold conditions to which it is adaptive but by day length. The switch from dissected submerged to floating entire leaves in *Ranunculus aquatilis* may not be brought about by submergence but by day length (Cook, 1968).

The reasons for indirect control systems can only be speculated upon. But it would appear in the case of leaf shedding in deciduous trees that if the shedding only took place when cold conditions occurred, the shedding response would be too late. As a result serious damage could occur which would outweigh the marginal gain from extended leaf retention. A reliable environmental signal occurring ahead of the potential stress factor is therefore used as the trigger, so that the response is complete by the time of onset of the stress factor. It is therefore interesting that where leaf shedding occurs in relation to irregularly occurring drought, as in cocoa (*Theobroma cacao*) and the curious desert shrub paloverde (*Cercidium floridum*), where there can be no early trigger, leaf fall is stimulated directly by drought (Alvim *et al.*, 1974). Direct response systems allow the degree of response to be related to the intensity of the stress. Nowhere is this more clear than in response to density where elegant relationships exist between density and plant weight (Harper, 1977). This type of response was termed dependent morphogenesis by Schmalhausen (1949).

Nevertheless, it is of interest that response systems involved in density are not necessarily straightforward. Plant height does not decrease with increasing crowding effects from other plants as does plant weight, but usually increases, for good adaptive reasons. These responses are brought about by a phytochrome system which responds not to amount but quality of light–red/far-red ratio (as shown in Table 5).

In this sort of analysis we tend to examine situations where facultative response systems are, or at least appear to be, producing responses which are adaptive. It must be realized that there are limits to what such systems can achieve. It is very difficult, for instance, to envisage a response system by which a plant growing in an exposed situation can become adapted almost instantaneously to a sudden severe gale. Animals can immediately move into protected parts of their habitat; plants have to sit the gale out. If exposure increased slowly they could respond by the sort of changes of growth form which many plants achieve under continuous exposure, particularly shortening of internodes and increase of branching, but they cannot do this quickly. Most species which extend their range into such exposed conditions therefore cope with such intermittent stress by the evolution of constitutive adaptations such as described by Turesson (1925) as *oecotypus campestris,* and well exemplified in the cliff populations of *Agrostis stolonifera* (Aston & Bradshaw, 1966) already referred to. The populations are effectively "sewn into their winter underwear" because they cannot change quickly enough (Bradshaw, 1965) and pay the penalty of reduced performance when the stress to which they are adapted does not occur. It is interesting that there are a very few species, notably *Mimosa pudica,* in which a very rapid response system has evolved.

This raises a final point. Evolution is, as Jacob (1977) has expressed so succinctly, a process of 'tinkering'. It has to make do with what is already there although mutation can add new variation very slowly. So the evolution of response systems will be substantially influenced by ancestry, as is all evolution. Thus the tillering response to the stress of grazing and trampling shown by the Gramineae is much related to their basic morphology which allows rapid regrowth from tillers. Nevertheless species in other families can also achieve the same response, so we should not exaggerate the limits set by ancestry. In this respect it is interesting that the phytoalexin response system is by no means limited to certain families or genera, but occurs widely. However the actual substances involved are quite different in different families (Table 8), which must reflect differences in ancestry (Harborne, 1982).

TABLE 8. Different types of phytoalexin produced in various plant families (Harborne, 1982)

Family	Phytoalexin type	Genera studied	Example
Amaryllidaceae	Flavan	*Narcissus*	7-Hydroxyflavan from daffodil, *N. pseudonarcissus*
Chenopodiaceae	Isoflavonoid	*Beta*	Betavulgarin from sugar beet *B. vulgaris*
Compositae	Polyacetylene	*Carthamus, Dahlia*	Safynol from safflower *C. tinctoria*
Convolvulaceae	Sesquiterpenoid furanolactone	*Ipomoea*	Ipomeamarone from sweet potato *I. batatas*
Euphorbiaceae	Diterpene	*Ricinus*	Casbene from castor bean *R. communis*
Gramineae	Diterpene	*Oryza*	Momilactone A from rice, *O. sativa*
Leguminosae	Isoflavonoids	Many	Pisatin from pea, *Pisum sativum*
Linaceae	Phenylpropanoid	*Linum*	Coniferyl alcohol from flax, *L. usitatissimum*
Malvaceae	Terpenoid polyphenol	*Gossypium*	Gossypol from cotton *G. barbadense*
Moraceae	Benzofuran	*Morus*	Moracin C from mulberry, *M. alba*
Orchidaceae	Hydroxyphenanthrene	*Loroglossum, Orchis*	Orchinol from military orchid, *O. militaris*
Rosaceae	Benzoic acids	*Malus, Prunus*	Benzoic acid from apple *M. pumila*
Solanaceae	Oxygenated sesquiterpenes	*Capsicum, Datura, Lycopersicon, Nicotiana, Solanum*	Rishitin from potato *S. tuberosum*
Umbelliferae	Isocoumarins	*Daucus*	6-Methoxymellein from carrot, *D. carota*
Vitaceae	Hydroxystilbene	*Vitis*	Resveratrol from grape vine, *V. vinifera*

CONCLUSIONS

We hope we have made clear that evolution is an almost inevitable outcome of stress. Because stress automatically gives rise to natural selection, evolutionary change will occur providing that the appropriate genetic variation is present. However, the nature of the change will depend very much on the nature of the stress factor and also the species. There will inevitably be great differences in detail—the mechanisms of tolerance to metal toxicity and to drought, for instance, are hardly likely to be the same. But there are more major differences, which relate to the pattern of occurrence of the stress, which mean that fundamentally different forms of adaptation are appropriate, either constitutive of facultative.

Even with this understanding we are not yet, however, in a position to explain or predict all evolution in relation to stress. Far too little is known about the range of variability available on which stress factors can act, particularly where facultative, plastic, responses are concerned. Although we can see how such response systems can be built up we know little about the building blocks for even one system.

The variability available must, of course, be appropriate. In this respect it will be inappropriate if it provides adaptation to the stress factor but entails too high costs in relation to other aspects of fitness. Evolution depends on a positive balance between costs and benefits. We know all too little about these even for constitutive adaptations, and even less for facultative adaptations. If these costs and benefits are known then predictions can be made (see Sibly & Calow, p. 101).

Then beyond this are other considerations which can lead to evolutionary inertia, particularly unfavourable linkage and pleiotropy. From all this it follows that we have still to answer, in relation to stress, all the questions put to physiological ecologists by Calow (1988). We can see very clearly that evolution in relation to stress does take place, and we can understand some of the guiding principles. We can also see we have still a long way to go to understand all that such evolution involves.

ACKNOWLEDGEMENTS

We are very grateful for the support of several colleagues in the preparation of this paper, particularly Julian Collins, Hamish Collin, David Thurman and George Russell, and Professor J. B. Harborne and Academic Press for permission to reproduce Table 8 in its entirety.

REFERENCES

AL-HIYALY, S. A., McNEILLY, T. & BRADSHAW, A. D., 1988. The effects of zinc contamination from electricity pylons—evolution in a replicated situation. *New Phytologist, 110:* 571–580.

ALLARD, R. W. & BRADSHAW, A. D., 1964. The implications of genotype-environment interactions in applied plant breeding. *Crop Science, 4:* 503–508.

ALVIM, R., ALVIM, P.de T., LORENZI, R. & SAUNDERS, P. F. 1974. The possible role of abscisic acid and cytokinins in growth rhythms of *Theobroma cacao* L., *Revista Theobroma, 4:* 3–12.

ANDERSON, J. M., 1986. Photoregulation of the composition, function, and structure of thylakoid membranes. *Annual Review of Plant Physiology, 37:* 93–136.

ANTONOVICS, J., BRADSHAW, A. D. & TURNER, R. G., 1971. Heavy-metal tolerance in plants. *Advances in Ecological Research, 7:* 1–85.

ARNTZEN, G. J., PFISTER, K. & STEINBACK, K. E., 1982. The mechanism of chloroplast triazine resistance: alterations in the site of herbicide action. In H. M. LeBaron & J. Gressel (Eds), *Herbicide Resistance in Plants:* 185–214. New York: Wiley.

ASHRAF, M., McNEILLY, T. & BRADSHAW, A. D., 1986a. The potential for evolution of salt (NaCl) tolerance in seven grass species. *New Phytologist, 103:* 299–309.

ASHRAF, M., McNEILLY, T. & BRADSHAW, A. D., 1986b. Tolerance of sodium chloride and its genetic basis in natural populations of four grass species. *New Phytologist, 103:* 725–734.

ASTON, J. L. & BRADSHAW, A. D., 1966. Evolution in closely adjacent plant populations. II. *Agrostis stolonifera* in maritime habitats. *Heredity, 21:* 649–664.

BAILEY, J. A., 1987. Phytoalexins: a genetic view of their significance. In P. R. Day & G. J. Jellis (Eds), *Genetics and Plant Pathogenesis:* 233–244. Oxford: Blackwell.

BAKER, A. J. M., 1987. Metal tolerance. *New Phytologist, 106* (suppl.): 93–111.

BJORKMAN, O. & HOLMGREN, P., 1963. Adaptability of the photosynthetic apparatus to light intensity in ecotypes from exposed and shaded habitats. *Physiologia Plantarum, 16:* 889–914.

BRADSHAW, A. D., 1965. Evolutionary significance of phenotypic plasticity in plants. *Advances in Genetics, 13:* 115–155.

BRADSHAW, A. D., 1973. Environment and phenotypic plasticity. *Brookhaven Symposium in Biology, 25:* 75–94.

BRADSHAW, A. D., 1984. The importance of evolutionary ideas in ecology—and vice versa. In B. Shorrocks (Ed.), *Evolutionary Ecology:* 1–25, Oxford: Blackwell.

BRADSHAW, A. D., 1987. Comparison—its scope and limits. *New Phytologist, 106* (suppl.): 3–21.

CALOW, P., 1988. Quo vadis? *Functional Ecology, 2:* 113–114.

COOK, C. D. K., 1968. Phenotypic plasticity with particular reference to three amphibious plant species. In V. H. Heywood (Ed), *Modern Methods in Plant Taxonomy:* 97–112. London: Academic Press.

COOK, S. A. & JOHNSON, M. P., 1968. Adaptation to heterogeneous environments, I. Variation in heterophylly in *Ranunculus flammula* L. *Evolution, 22:* 449–516.

COOPER, J. P., 1954. Studies on growth and development in *Lolium.* IV. Genetic control of heading responses in local populations. *Journal of Ecology, 42:* 521–556.

COOPER, J. P., 1964. Climatic variation in forage grasses: Leaf development in climatic races of *Lolium* and *Dactylis. Journal of Applied Ecology, 1:* 45–61.

DARVILL, A. G. & ALBERSHEIM, P., 1984. Phytoalexins and their elicitors—a defence against microbial infection in plants. *Annual Review of Plant Physiology, 35:* 243–275.

ERNST, W., 1976. Physiological and biochemical aspects of metal tolerance. In T. A. Mansfield (Ed.), *Effects of Air Pollutants on Plants:* 115–133. London: Cambridge University Press.

FINLAY, K. W. & WILKINSON, G. N., 1963. The analysis of adaptation in a plant breeding programme. *Australian Journal of Agricultural Research, 14:* 742–754.

GARTSIDE, D. W. & McNEILLY, T., 1974. Genetic studies in heavy metal tolerant plants. III. Zinc tolerance in *Agrostis tenuis. Heredity, London, 33:* 303–308.

GRIME, J. P., 1977. Evidence for the existence of three primary strategies in plants and its relevance to ecological and evolutionary theory. *American Naturalist, 111:* 1169–1194.

GRIME, J. P., CRICK, J. C. & RINCON, J. E., 1986. The ecological significance of plasticity. In D. H. Jennings & A. J. Trewavas (Eds), *Plasticity in Plants:* 5–29. Cambridge: Company of Biologists.

HARBORNE, J. B., 1982. *Introduction to Ecological Biochemistry.* London: Academic Press.

HARPER, J. L., 1977. *Population Biology of Plants.* London: Academic Press.

JACOB, F., 1977. Evolution and tinkering. *Science, 196:* 1161–1166.

JAIN, S. K., 1978. Inheritance of phenotypic plasticity in soft chess, *Bromus mollis* L. (Gramineae). *Experientia, 34:* 835–836.

JENNINGS, D. H. & TREWAVAS, A. J. (Eds), 1986. *Plasticity in Plants.* Cambridge: Company of Biologists.

JINKS, J. L. & MATHER, K., 1955. Stability in development of heterozygotes and homozygotes. *Proceedings of the Royal Society, Series B, 143:* 561–578.

KERNER VON MARILAUN, A. & OLIVER, F. W., 1895. *The Natural History of Plants, Vol. 2.* Blackie, London.

KHAN, M. A., ANTONOVICS, J. & BRADSHAW, A. D., 1976. Adaptation to heterogeneous environments, III. The inheritance of response to spacing in flax and linseed (*Linum usitatissimum*). *Australian Journal of Agricultural Research, 217:* 649–659.

KHAN, M. I., 1967. *The Genetic Control of Canalisation of Seed Size in Plants.* Ph.D. thesis, University of Wales.

LANG, A. R. G. & BEGG, J. E., 1979. Movements of *Helianthus annus* leaves and heads. *Journal of Applied Ecology, 16:* 299–306.

LARCHER, W. & BAUER, H., 1981. Ecological significance of resistance to low temperature. In O. L. Lange, P. S. Nobel, C. B. Osmond & H. Ziegler (Eds), *Encyclopaedia of Plant Physiology, New Series, vol. 12A:* 403–437. Berlin: Springer-Verlag.

LAWRENCE, M. J., 1984. The genetical analysis of ecological traits. In B. Shorrocks (Ed.), *Evolutionary Ecology:* 27–63. Oxford: Blackwell.

LeBARON, H. M. & GRESSEL, J., 1982. *Herbicide Resistance in Plants.* New York: Wiley.

LICHTENTHALER, H. K., 1985. Differences in morphology and chemical composition of leaves grown at different light intensities and qualities. In N. R. Baker, W. J. Davies & C. K. Ong (Eds), *Control of Leaf Growth:* 203–221. Cambridge: Cambridge University Press.

LOSCH, R. & TENHUNEN, J. D., 1981. Stomatal responses to humidity. In P. G. Jarvis & T. A. Mansfield (Eds), *Stomatal Physiology:* 137–161. Cambridge: Cambridge University Press.

MACNAIR, M. R., 1981. Tolerance of higher plants to toxic materials. In J. A. Bishop & L. M. Cook (Eds), *Genetic Consequences of Man-Made Change:* 177–207. London: Academic Press.

McNEILLY, T., ASHRAF, M. & VELTKAMP, C., 1987. Leaf micromorphology of sea cliff and inland plants of *Agrostis stolonifera* L., *Dactylis glomerata* L. and *Holcus lanatus* L. *New Phytologist, 106:* 261–269.

MATSUO, T. (Ed), 1975. *Adaptability in Plants—with Special Reference to Crop Yield.* Japan: University of Tokyo.

MOONEY, H. A. & GULMON, S. L., 1979. Environmental and evolutionary constraints on the photosynthetic characteristics of higher plants. In O. T. Solbrig, S. K. Jain, G. B. Johnson & P. H. Raven (Eds), *Topics in Plant Population Biology:* 316–337. London: Macmillan.

MORGAN, D. C., 1981. Shadelight quality effects on plant growth. In H. Smith (Ed), *Plants and the Daylight Spectrum:* 205–221. London: Academic Press.

PERKINS, J. M. & JINKS, J. L., 1973. The assessment and specificity of environmental and genotype-environmental components of variability. *Heredity, 30:* 111–126.

SCHMALHAUSEN, I. I., 1949. *Factors in Evolution.* New York: McGraw-Hill.

SCHLICHTING, C. D., 1986. Evolution of phenotypic plasticity in plants. *Annual Review of Ecology and Systematics, 17:* 667–693.

SLADE, A. J. & HUTCHINGS, M. J., 1987. Clonal integration and plasticity in foraging behaviour in *Glechoma hederacea. Journal of Ecology, 75:* 1023–1036.

SMITH, H. (Ed), 1981. *Plants and the Daylight Spectrum.* London: Academic Press.

SNAYDON, R. W. & BRADSHAW, A. D., 1961. Differential responses to calcium within the species *Festuca ovina* L. *New Phytologist, 60:* 219–234.

SØRENSEN, T., 1954. Adaptation of small plants to deficient nutrition and a short growing season. Illustrated by cultivar experiments with *Capsella bursa-pastoris* (L). *Botanisk Tidsskrift, 51:* 339–361.

SULTAN, S. E., 1987. Evolutionary implications of phenotypic plasticity in plants. *Evolutionary Biology, 21:* 127–178.

TUCKER, J. M., 1952. Evolution in the Californian Oak *Quercus alvordiana. Evolution, 6:* 162–180.

TURESSON, G., 1925. The plant species in relation to habitat and climate. *Hereditas, 6:* 147–234.

VIRGIN, H. I., 1964. Some effects of light on chloroplasts and plant protoplasm. In A. C. Giese (Ed.), *Photophysiology:* 273–303. New York: Academic Press.

WADDINGTON, C. H., 1960. Experiments on canalising selection. *Genetical Research, 1:* 140–150.

WALLEY, K. A., KHAN, M. S. I. & BRADSHAW, A. D., 1974. The potential for evolution of heavy metal tolerance in plants. I. Copper and zinc tolerance in *Agrostis tenuis. Heredity, London, 32:* 309–319.

WESTCOTT, B., 1986. Some methods of measuring genotype-environment interaction. *Heredity, London, 56:* 243–254.

WILLIAMS, W., 1960. Relative variability of inbred lines and F_1 hybrids in *Lycopersicum esculentum. Genetics, 45:* 1457–1465.

WOOLHOUSE, H. W., 1983. Toxicity and tolerance in the response of plants to metals. In O. L. Lange, P. S. Nobel, C. B. Osmond & H. Ziegler (Eds), *Encyclopaedia of Plant Physiology New Series, vol. 12C:* 245–300. Berlin: Springer-Verlag.

WU, L., BRADSHAW, A. D. & THURMAN, D. A., 1975. The potential for evolution of heavy metal tolerance in plants. III. The rapid evolution of copper tolerance in *Agrostis stolonifera. Heredity, London, 34:* 165–187.

YOUNG, A. J., COLLINS J. C. & RUSSELL, G., 1987. Ecotypic variation in the osmotic responses of *Enteromorpha intestinalis* (L). Link. *Journal of Experimental Botany, 38:* 1309–1324.

Biological Journal of the Linnean Society (1989), *37:* 157–171. With 6 figures

Towards a physiological and genetical understanding of the energetics of the stress response

RICHARD K. KOEHN

Department of Ecology and Evolution, State University of New York, Stony Brook, New York 11794, U.S.A.

AND

BRIAN L. BAYNE

Plymouth Marine Laboratory, Prospect Place, The Hoe, Plymouth PL1 3DH

We consider stress as an environmental change that results in reduction of net energy balance (i.e. growth and reproduction). Reduced energy balance restricts the environmental range of an organism and may change the environmental optima at which maximum production can be achieved. We emphasize individual differences in net energy balance and the interrelationships among genetic heterozygosity, rate of protein synthesis, efficiency of protein synthesis and whole organism measures of both routine and maintenance metabolic rate. Lastly, we consider the consequences of genetically determined individual differences in metabolic maintenance costs within the context of variable environments and how genetic/environmental interactions can define individual responses to environmental extremes.

KEY WORDS:—Stress – energy balance – genetic factors –heterozygosity.

CONTENTS

INTRODUCTION

In this paper we shall consider the effects of stress primarily from the point of view of the physiological energetics of the organism; i.e. the physiological traits (feeding, absorption and the metabolic costs of maintenance, growth and reproduction) that together comprise the energy budget. We consider that any

0024–4066/89/050157+15 $03.00/0

reduction in production (somatic growth, reproduction or both) in response to an environmental change signifies reduced Darwinian fitness, and therefore represents a result of environmental stress. Individual organisms differ in energy balance with respect to the presence and the absence of apparent environmental sources of stress; hence the role of genotype in the determination of individual energy balance and the response to stress will be considered. Lastly, we shall consider some of the mechanisms by which genetic and physiological variability may be related.

Stress may be defined as any environmental change that acts to reduce the fitness of an organism. This is a general restatement of the definition offered by Brett (1958) that stress refers to "any environmental factor which extends the normal adaptive response of an animal, or which disturbs the normal functioning to such an extent that the chances of survival are significantly reduced". In any attempt to measure the response of an organism to such stresses, three general considerations should be borne in mind. First, the effects of the stress will be an integrated response involving all levels of functional complexity within the organism (molecular, cellular and physiological). Secondly, the stress response is dynamic, and involves an alteration in functional properties over time. And thirdly, a potential stress may be neutralized by homeostatic physiological compensation, although these processes may themselves be metabolically costly. It is when compensation for an environmental change is incomplete or, in the extreme, impossible, that lasting effects are measurable as a decline in the organism's fitness or, ultimately, as death.

For practical convenience, an organism's response to a potential stress may be viewed on different time scales, with an implied cascade of cause and effect. Responses have been usefully categorized as primary, secondary and tertiary effects (Selye, 1950). For example, from research on fishes (Wedemeyer & McLeay, 1981): primary effects include release of hormones from endocrine tissue; secondary effects include changes in blood chemistry, depletion of nutrient stores and metabolic changes such as negative nitrogen balance; and tertiary effects involve impaired growth and reproduction and increased vulnerability to disease.

We shall be concentrating on marine molluscs, where there is a less detailed understanding of the primary through tertiary responses. Nevertheless, some features of an integrated response to environmental stress have emerged from recent research. For example, Moore *et al.* (1987) have described cellular and tissue responses in the bivalve *Mytilus edulis* to stress from petroleum hydrocarbons. They suggested causal links between alterations to the functional properties of lysosomal membranes, cellular autolysis and changes in the proportional distribution of cell types within digestive and reproductive tissues, with ensuing consequences for growth and reproduction. Moore & Viarengo (1987) demonstrated a causal link between lysosomal destabilization and increased protein catabolism in *Mytilus* digestive cells. As we shall discuss later, protein catabolism, leading to more rapid turnover (breakdown and synthesis) of tissue proteins, reflects an increase in the costs of maintenance (Hawkins, 1985) with consequences for whole-animal energy balance and growth. Widdows, Donkin & Evans (1987) have demonstrated dose-response relationships between exposure to hydrocarbons and the growth of *Mytilus*.

STRESS AND PHYSIOLOGICAL ENERGETICS

The balance between energy gains and losses to an organism may be represented in the "balanced energy equation" (Winberg, 1956). For the purpose of this paper we write the energy budget as follows (Bayne & Newell, 1983):

$$Pg + Pr = C*AE - (R_m + R_r)$$

where Pg is somatic production, Pr is reproductive (gamete) production, C is energy consumption, AE is the efficiency with which consumed energy is absorbed by the animal, R_m represents the metabolic costs of body maintenance (measured as steady-state oxygen consumption in the absence of food) and R_r includes all other metabolic costs (feeding, growth and reproduction, as measured during feeding). For practical purposes, R_r is determined as $R_t - R_m$, where R_t is the total measured metabolic rate. This formulation of the energy budget ignores losses due to excretion (which are small in most marine bivalve molluscs; Bayne & Newell, 1983). (C*AE) represents the absorbed ration (A), and (Pg + Pr) may be referred to as the "scope for growth and reproduction" (SFG), "energy balance" or "production".

An environmental change represents a stress when the balance between A and $(R_m + R_r)$ is disturbed with consequent reduction in (Pg + Pr). An organism may normally be able to maintain SFG constant over a range of values of an environmental variable, failing to do so (i.e. experiencing stress) only at the extremes of the normal range. For example, Bayne, Thompson & Widdows (1973) measured the SFG of mussels, *Mytilus edulis*, in steady-state conditions at various temperatures (Fig. 1) and distinguished between a "zone of tolerance" (Fry, 1947), within which the effects of temperature were minimal, and a zone of stress within which temperature had a marked and deleterious effect on scope for growth. Stickle & Bayne (1987) measured SFG of the dogwhelk, *Nucella lapillus*, as a function of temperature and salinity and described, from these laboratory experiments, an environmental range over which SFG was negative, signifying lasting stress; this range was shown to match the known limits to the distribution of dogwhelks in estuaries and on the shore.

Such physiological responses to an environmental change are, however, dynamic; the initial physiological imbalance caused by the change may be overcome by physiological compensation (acclimation), resulting in a transient stress effect only. Figure 2 shows the acclimation of *Mytilus edulis* to a change in the quantity and the quality of the food (Bayne, Hawkins & Navarro, 1987); on first experiencing the new rations, SFG was depressed to negative values, but modifications to feeding behaviour and to total metabolic rate over a period of two weeks effected an increase in SFG and a full compensation for differences in food quality (measured here as the mass of organic matter available per unit volume of suspended particulate material), so neutralizing the potential stress from the change in diet.

The net effect of such adjustments between the physiological components of production, then, is to effect an apparent homeostatic control over growth, buffering the allocation of energy to growth and reproduction against changes in an environmental variable. In the natural environment the organism is affected by many such variables acting together. In such circumstances the effects of

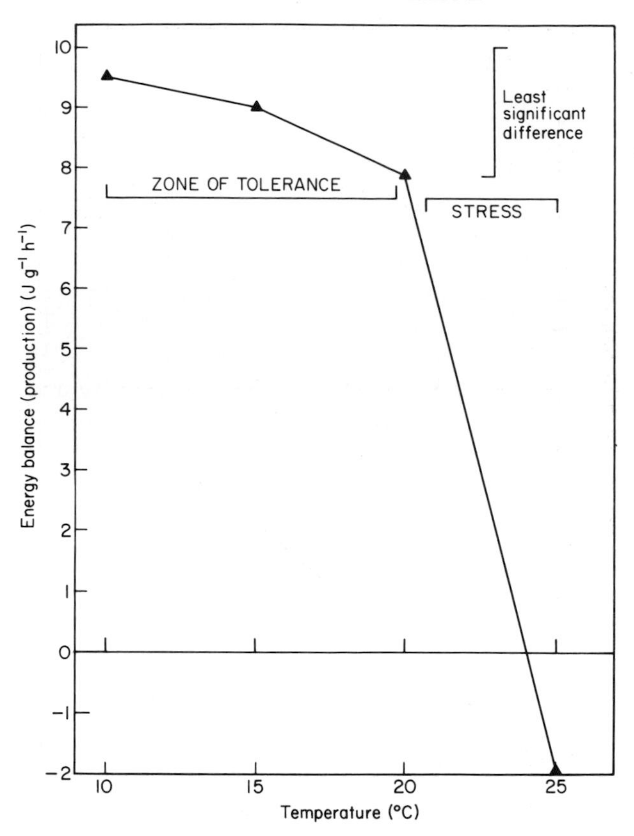

Figure 1. The energy balance of *Mytilus edulis* at different temperatures. Between 10 and 20°C physiological compensations result in relatively constant scope for growth; between 20 and 25°C, physiological compensation is impaired signalling the effects of stress. (From Bayne, Thompson & Widdows, 1973.)

stress may be to reduce the environmental range over which the organism may maintain positive values for production, and also to alter the environmental optimum at which production is maximized.

For example, Fig. 3 describes absorbed ration (A) and routine metabolic rate (R_r) in *M. edulis* as functions of the availability of food, measured as the concentration of particulate organic matter (POM) per litre (Bayne *et al.*, 1988). As a result of a postulated environmental change which increases R_m by 25% (e.g., R_{t1} to R_{t2}; Fig. 3), not only will SFG (as $A-R_t$) be reduced overall, but the minimum concentration of food at which production can be maintained positive will be increased (A *vs.* B; Fig. 3); physiological flexibility with respect to particulate organic matter in suspension is reduced, with potentially damaging effect on overall fitness.

Bayne (1985) considered the effects of temperature stress on the thermal optimum for growth of *M. edulis*. Absorbed ration (A) increases with rise in temperature to 15°C, as a result of increased feeding rate, then declines as filtration rate is suppressed at higher temperatures (Fig. 4). Total respiration rate, however, increases in a more complex manner with rise in temperature; the result is a maximum SFG (A-R_{t1}; Fig. 5) at 14°C. Should routine metabolic rate

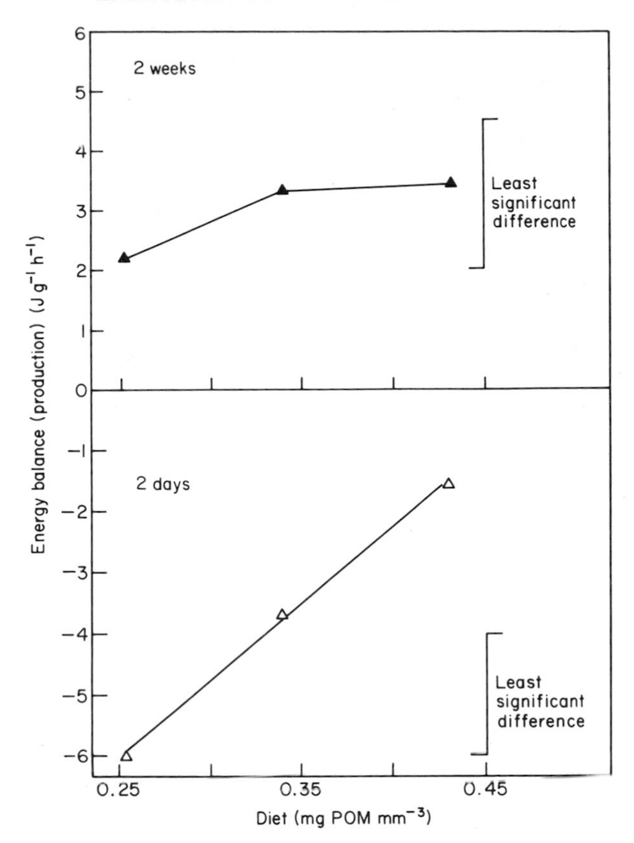

Figure 2. The energy balance of *Mytilus edulis* held for two days and two weeks on diets of different nutritional quality (as mg organic matter–POM–per mm^3 of particulate material). Physiological compensation over two weeks increases SFG, overcoming the initial stress imposed by the experimental diets. (From Bayne, Hawkins & Navarro, 1987.)

now be increased (e.g. as a result of a change in salinity), differences between the thermal dependence of R_m and R_r result in a SFG which now shows a maximum (and therefore a thermal optimum) at 11.5°C (A-R_{t2}; Fig. 5). For an organism synchronizing a seasonal reproductive cycle to temperature, such a shift in the thermal optimum may have serious implications for breeding success.

When viewed in the context of physiological energetics, therefore, stress acts to constrain growth and/or reproduction, reducing what Fry (1947) called the "zone of tolerance", limiting the range of environment over which the organism may, by balancing physiological traits, maximize growth potential, and possibly also modifying environmental optima.

CAUSAL FACTORS IN VARIABILITY IN PHYSIOLOGICAL RESPONSES TO STRESS

Individual variability and performance

Somatic production, or growth, is therefore a reflection of the different elements of the balanced energy equation that collectively represent both the supply and demand of energy for an individual. Individual growth rate reflects

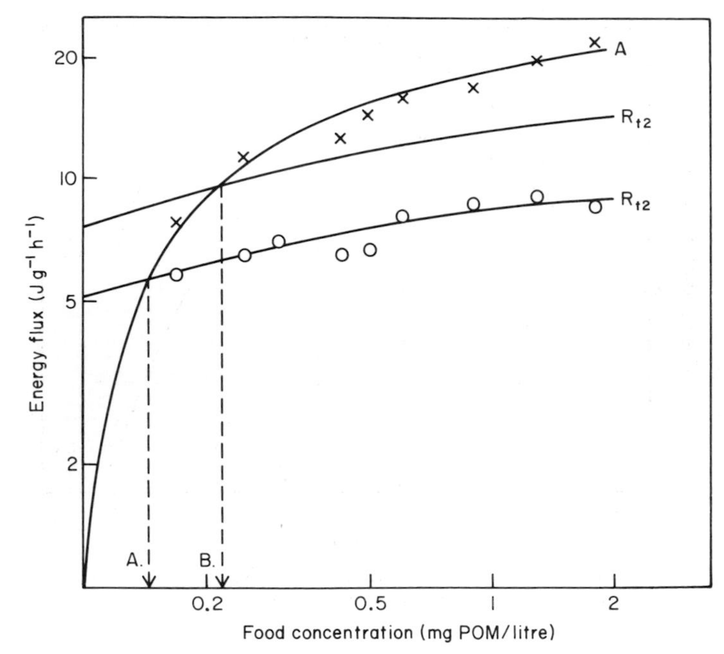

Figure 3. The absorbed ration (A: Joules g^{-1} h^{-1}) and respiration rates (R: J g^{-1} h^{-1}) for *Mytilus edulis* at different concentrations of food. R_{t2} represents a 25% increase in respiration rate over R_{t1} and results in a higher food threshold (B cp. A) below which SFG (as $A-R_t$) is negative. (From Bayne *et al.*, 1988.)

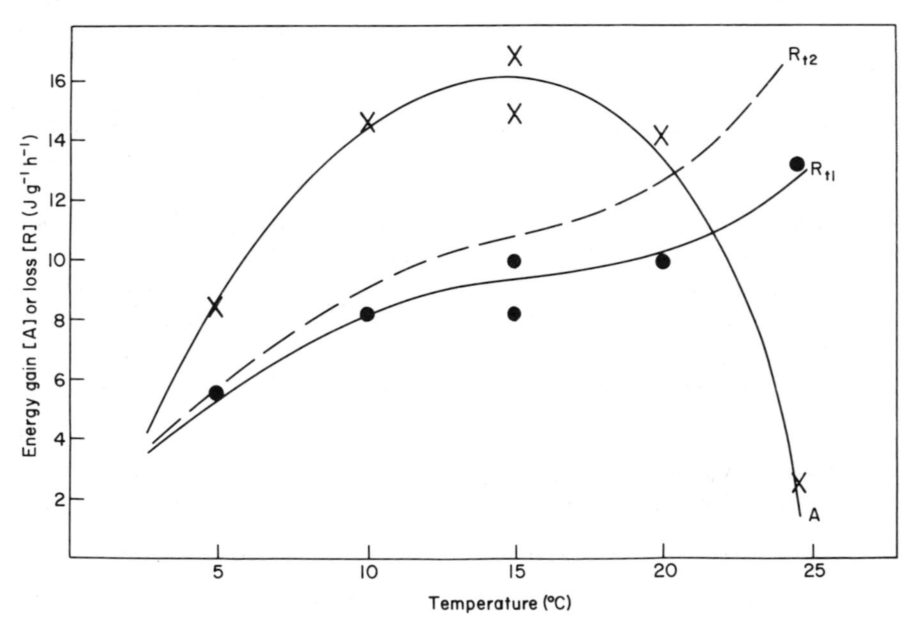

Figure 4. Energy gain (A: J g^{-1} h^{-1}) and loss (R: J g^{-1} h^{-1}) by *Mytilus edulis* at different temperatures. The change in total metabolic rate is non-linear. R_{t1} and R_{t2} represent differences in metabolic rate caused by an environmental change or by genetic differences between individuals. (From Bayne, 1985.)

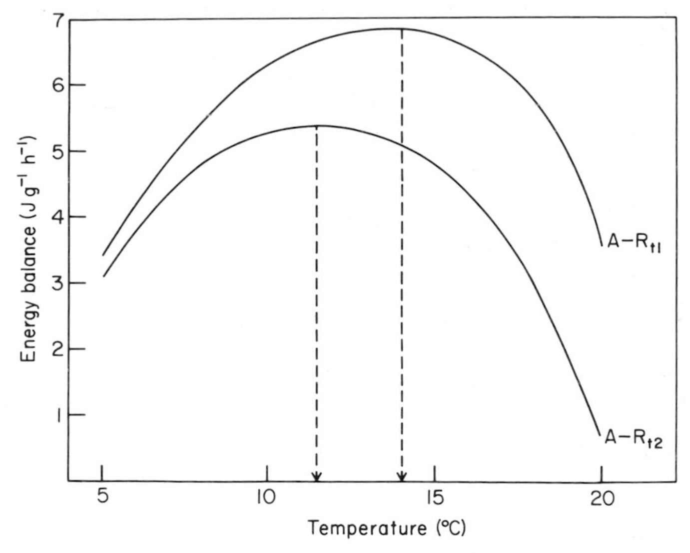

Figure 5. Energy balance ($A \times R_t$) computed from R_{t1} and R_{t2} (Fig. 4) to illustrate the shift in thermal optimum (dashed arrows) that can result from changes in R_t.

the summation of the various terms that can be either negative, positive or zero. However, the relative contribution of each term cannot be assessed from a measure of growth rate alone; variability in individual growth rate can potentially derive from individual differences in R_m, R_r, and/or consumption (C).

Table 1 summarizes measured degrees of variability (in *Mytilus*) for rates of consumption (C; mg organic matter hour^{-1}) and maintenance and routine respiration rates (ml O_2 consumed hour^{-1}). All values have been corrected for differences in body size (as dry flesh weight for individuals of similar shell length) though age differences (see later) and measurement errors (which are small relative to overall variance) are still represented; coefficients of variation are between 23 and 33%.

Hawkins, Bayne & Day (1986) have attempted to analyse the causes of such variability in maintenance metabolic rate and to relate this to variability in other physiological traits, including growth. Table 2 shows relationships between the energy demands for maintenance (R_m), net protein balance (i.e. growth in terms of protein, P), protein synthesis ($\mathcal{Z}$) and the efficiency of protein synthesis ($P/\mathcal{Z}$) for mussels from a single cohort. Within this cohort, individuals grew at

TABLE 1. Variability in three physiological rates (*Mytilus edulis*). Sample size in each case was ten individuals of similar shell length, and rates are standardized 1 g dry flesh weight

Physiological rate	Mean	Standard deviation
Consumption (mg organic matter h^{-1})	1.90	0.63
Maintenance respiration (ml O_2 h^{-1})	0.348	0.112
Routine respiration (ml O_2 h^{-1})	0.264	0.061

TABLE 2. Aspects of protein synthesis and maintenance metabolism (R_m) in two groups of mussels (*Mytilus edulis*) of the same age, one slow growing and one fast growing group. Values are means ± 2 standard errors, for n = 4 replicates, each of > 5 mussels. Protein balance represents protein absorbed from the food minus protein lost through excretion; protein synthesis is gross synthetic rate. Data from Hawkins, Bayne & Day, 1986

Experimental group	R_m ($J\ d^{-1}$)	Protein balance ($mg\ d^{-1}$)	Protein synthesis ($mg\ d^{-1}$)	Efficiency (%) of synthesis
Slow	73.4	1.19 ± 0.30	3.23 ± 0.53	36.8 ± 5.4
Fast	47.4	1.93 ± 0.38	2.75 ± 0.58	70.2 ± 4.3

different rates and two groups were identified for measurements after five months, one of slow growing mussels (mean dry flesh weight $\pm 2 \times$ S.E. $= 7.61 \pm 1.06$ mg), and the other of faster growing individuals (60.7 ± 6.71 mg). All values were standardized to an equivalent body mass of 60.7 mg.

Slow growth was marked by high rate of maintenance metabolism, a high rate of protein systhesis and a low synthetic efficiency, relative to the faster growing group. Hawkins *et al.* (1986) concluded that protein synthesis, comprising a major element of maintenance metabolism, carried a high metabolic cost (representing at least 20% of total energy expenditure; Hawkins, Widdows & Bayne, 1988); individuals with reduced efficiency of protein synthesis therefore incurred a higher metabolic expenditure on maintenance and a lower net rate of somatic growth.

In summary, individual differences in growth derive from differences in energy balance. Individual energy balance reflects, in part, differing costs of metabolic maintenance, caused by variability in the efficiency of protein synthesis. The existence of individual variability in energy metabolism suggests the possibility that these differences are due, at least in part, to genotype.

Genetic correlates of individual variability in energy balance and performance

Somatic production (Pg), or scope for growth, is a direct reflection of net energy balance. That Pg has a genetic component has been recognized for years, and a large literature on animal and plant breeding supports this point. Unfortunately, the selective breeding of organisms for increased (or decreased) growth rate, while establishing a genetic determinant for growth rate, does not provide sufficiently precise information to formulate a genetic model for growth. Only recently have specific genetic correlates of growth become apparent.

Correlations between the degree of individual herterozygosity, measured at a series of electrophoretically detectable enzyme genes, and various age and/or growth rate parameters, have been established for a wide diversity of plant and animal species (Mitton & Grant, 1984; Zouros & Foltz, 1987). Marine invertebrates have been particularly well studied, including the gastropod *Thais haemastoma* (Garton, 1984), the bivalves *Crassostrea virginica* (Singh & Zouros, 1978, 1981; Zouros, Singh & Miles, 1980), *Crassostrea gigas* (Fujio, 1982), *Mulinia lateralis* (Garton, Koehn & Scott, 1984; Koehn, Diehl & Scott, 1988), *Macoma baltica* (Green *et al.*, 1983), and *Mytilus edulis* (Koehn & Gaffney, 1984; Diehl & Koehn, 1985). [Readers interested in other plant and animal species should consult Bush, Smouse & Ledig (1987) and Danzmann & Ferguson (1988).] Thus, there is a substantial body of literature, from a diversity of organisms, demonstrating that genetic heterozygosity is positively and significantly correlated with rates of somatic growth among individuals in outbred populations. Genetic and/or environmental circumstances that militate against heterozygosity-dependent growth rate have been discussed by Gaffney & Scott (1984) and Koehn *et al.* (1988).

Although the precise mechanism by which heterozygosity leads to higher growth rate is presently unknown, these genetic effects on growth are apparently a consequence of heterozygosity-dependent R_m. Data on oxygen consumption, reflecting rates of maintenance metabolism, have been shown to be lower in more heterozygous individuals of oysters (Koehn & Shumway, 1982), mussels (Diehl *et al.*, 1985; Diehl, Gaffney & Koehn, 1986) a salamander (Mitton, Carey & Kocher, 1986), and a salmonid fish (Danzmann, Ferguson & Allendorf, 1988). Hawkins *et al.* (1986), while confirming the result of Koehn & Gaffney (1984) that more heterozygous individuals achieve faster growth rates, also determined that genotype-dependent energy balance derives from differences in protein turnover as a primary factor effecting both maintenance metabolic expenditure (Table 2) and consumption. Energy 'saved' in maintenance was expended to increase consumption (i.e. filtration) rate. Thus, it is not unexpected that consumption rates can also be correlated with heterozygosity, as was found in *Rangia cuneata* (Holley & Foltz, 1987).

If genetic heterozygosity is correlated with Pg, C, R_m, and the efficiency of protein synthesis, what is the functional connection between these measures of physiological and metabolic performance and genetic heterozygosity? Why should the genotype at a series of randomly selected enzyme loci have phenotypic effects on metabolism? Apparently, not all loci have metabolic effects. Rather, the metabolic effects of genetic heterozygosity derive only from variation at loci involved in ATP production and protein catabolism, the energetic supply and demand of protein turnover. Koehn *et al.* (1988) assessed the relative attribution among 15 loci of heterozygosity to growth rate in *Mulinia lateralis*. Loci with large and significant correlations with growth rate synthesized enzymes that function in protein catabolism or glycolysis; heterozygosity in enzymes of the pentose shunt, redox balance, detoxification, digestion, and other metabolic roles were not correlated with growth rate.

We conclude that individual performance in R_m, growth rate and efficiency of protein synthesis can be caused by the degree of individual heterozygosity of enzymes that support metabolic pathways of ATP production and protein catabolism. While the precise biochemical mechanisms by which heterozygosity effects metabolic efficiency are still unknown (but see Koehn *et al.*, 1988), for reasons described earlier relating maintenance costs to the stress response, we can predict that differences in the response to environmental stress will also be genotype dependent and will be linked with differences in maintenance metabolism.

Genetic variability and the effects of stress

The experiment of Hawkins *et al.* (1986) confirmed genotype-dependent protein turnover (i.e. the balance between protein synthesis and breakdown) as a primary factor affecting maintenance metabolic energy demand, and, therefore, as a contributor to variability between individuals in the rate of growth. *A priori*, high rates of protein turnover, although incurring high metabolic costs, may be expected to give an advantage under conditions of stress, where the rate of protein renewal may be limiting. For example, protein synthesis is a fundamental component of physiological acclimation to change in temperature (Hazel & Prosser, 1974), so a high inherent rate of protein synthesis might confer a fitness advantage under temperature stress.

Hawkins, Wilson & Bayne (1987) addressed this question by measuring rates of protein synthesis and breakdown, and components of the energy budget, for mussels subjected to increases in temperature above 10°C. Individuals expressing high rates of protein turnover were more sensitive to change in temperature, and more susceptible to ultimately lethal temperature. This study confirmed a negative relationship between rate and efficiency of protein synthesis and a positive relationship between the absolute changes in total oxygen consumption rate (R_t) and protein synthesis when the temperature was increased by 10°C, and between efficiency of protein synthesis and time to death at 28.5°C. In the case of temperature stress, therefore, higher rates of protein systhesis did not confer any advantage and, on the contrary appeared, by elevating metabolic demand, to result in a significant disadvantage compared with individuals with a lower and more efficient rate of protein synthesis.

We suggest that stress will act differentially on individuals according to the efficiency with which they meet their metabolic requirements (i.e. R_m as a fraction of R_t); the lower this fraction, the more surplus energy will be available, given invariant energy inputs, for growth and for resisting environmental stress. When energy resources in the environment are ample, genotype-dependent variability in maintenance efficiency may be a quantitatively minor factor in affecting individual differences in the rate of growth. As the environment becomes more stressful, by reducing the supply of energy (or nutrients), or by increasing the organism's metabolic demands, differences in maintenance efficiency will play a greater role in determining variability between individuals in their stress tolerance. It follows that correlations between genetic variability and physiological performance will be more apparent under conditions of potential stress than under more optimal conditions for growth and survival.

Several authors have suggested that genetic correlates of growth rate might be influenced by the degree to which organisms are subject to environmental perturbations or stress (Koehn & Shumway, 1982; Gaffney, 1986; Diehl *et al.*, 1986). Unfortunately, there are limited data available from specific tests of this prediction. Two recent experiments were specifically designed to test this point.

Individual juveniles of the bivalve *Mulinia lateralis* (Scott & Koehn, 1988) were collected and randomly subdivided into four groups. Each group was grown for a period of two weeks in the laboratory under differing conditions of temperature and salinity stress (Table 3). One group was a control, grown at constant conditions of 20°C and 22‰ salinity. The growth of the control group was approximately 50% greater than any of the three treatment groups, which did not differ in laboratory growth rate from one another. There was a significant correlation between heterozygosity, measured among twelve loci, and size at collection though, within each group, no significant relationships could be detected (Table 3), probably due to limited sample size (average group n=60).

TABLE 3. The effect of environmental stress on the relationship between heterozygosity at 12 loci and growth in *Mulinia lateralis*. Groups are: 1, control, constant 20°C and 22‰; 2, constant 20°C, salinity varied between 18‰ and 26‰ every 2 days; 3, constant 28°C and constant 22‰; and 4, constant 28°C and fluctuating salinity as in 2. All groups fed well above the maintenance requirement; sample size was, on average, 60 individuals per group. (From Scott & Koehn, 1988)

	[1]		[2]	
Group	r^2	*P*	r^2	*P*
1	0.066	0.477	0.044	0.554
2	0.012	0.355	0.061	0.021
3	0.013	0.457	0.163	0.006
4	0.021	0.232	0.008	0.444

[1] Coefficient of determination and associated statistical probability from regression of heterozygosity on initial size, at time of collection.
[2] Coefficient of determination and associated statistical probability from regression of heterozygosity on growth added during two weeks in the laboratory under specified conditions.

A highly significant relationship between heterozygosity and added growth in the laboratory was detected in two of the three experimental treatments. (Table 3). Varying salinity and elevated temperature each resulted in genotype-dependent growth; more heterozygous individuals achieved more growth than less heterozygous individuals. Surprisingly, there was no genotype-dependent growth among individuals experiencing both elevated temperature *and* fluctuating salinity. This may have been due to either the small sample size of the individual treatments, or to the combined effects of two different stress factors not being additive. Mortality of group 4 was lower than in group 3, further suggesting lower stress in group 4. In any event, these data suggest genotype/environmental interactions by which the increased energy demands that result from environmental stress enhanced the differential growth rate among genotypes.

Diehl (1988) tested for the effect of heterozygosity at 13 loci encoding 10 enzymes of carbohydrate metabolism on growth under limited food, low moisture stress in the juvenile manure worm *Eisenia foetida*. There was no relationship between individual heterozygosity and either initial weight or weight added during stress or non-stress treatments, but significant locus specific heterozygosity contributed to weight added during the stress (but not the non-stressed) treatment. These data, like those for *Mulinia* described above, suggest that the phenotypic effects of heterozygosity on growth are enhanced under environmental conditions that reduce net energy balance.

CONCLUSIONS

A genetic basis for the energy demands of maintenance had been demonstrated in animal husbandry (cf. Van Steenbergen, 1987) and the lower weight-specific rates of whole-animal metabolic rate seem to derive from individual differences in maintenance cost. Moreover, the individual costs of metabolic maintenance, reflected in R_m, derive from differences in the efficiency of protein metabolism and net protein balance. What is the relationship between genetically-determined individual differences in maintenance metabolism and individual differences in stress-related physiological performance? The effect of environmental change and/or stress upon individual variability in physiological performance is an area still in need of investigation. Nevertheless, we believe that on the basis of the data we have presented, we can formulate a hypothesis that addresses this question.

From Fig. 3, and interpreting R_{t1} and R_{t2} as representing two individuals with different levels of maintenance metabolism, two features are apparent. First, the two individuals will differ in net energy available to growth at all levels of absorption. Second, because of differences in maintenance costs, the point of zero scope for growth is also different between the two individuals, with animal 'A' enjoying a net positive energy balance at low levels of absorption where animal 'B' is in negative balance. We may also postulate that, as food increases to higher levels, differences in maintenance metabolism among individuals become smaller, relative to overall net energy balance. Conversely, as absorption reduces, the putative genetic component to variability in maintenance costs becomes more important; stress from reduced ration may therefore enhance the measurable relationship between genotype and phenotype (i.e. growth). These relationships will be complicated, of course, by the metabolic costs of growth itself, making empirical confirmation of causality between heterozygosity and growth more difficult as the net energy available for growth increases.

Similar considerations apply with respect to temperature (Figs 4 & 5). To the extent that high genetic heterozygosity is correlated with low maintenance cost, more heterozygous individuals are expected to exhibit positive growth over a broader temperature range than less heterozygous individuals (Koehn & Shumway, 1982; Hawkins *et al.*, 1987). At the extremes of the tolerable temperature range, differences between individuals in maintenance costs are likely to be larger proportions of the differences in net energy balance than at the temperature optimum for growth. Hence temperature stress should enhance the differences between individuals in genotype dependent rates of growth.

Factors other than temperature can act as a source of stress. For example, positive SFG occurs over a range of environmental salinity, but negative SFG will occur at both the upper and lower extremes of salinity because of increased maintenance costs. This is different from the case of temperature where costs of maintenance increase monotonically. Nevertheless, lower maintenance allows greater SFG at the optimum environment (i.e. salinity) as well as positive SFG over a greater environmental range (Fig. 6).

The cost of maintenance metabolism is genetically determined. When individual differences in maintenance costs occur, individuals will differ in the range of environments over which each may maintain positive SFG. This is true for very different sources of environmental stress. Moreover, in a variable environment, the variance among individuals in growth rate would be greater for individuals experiencing higher costs of metabolic maintenance than those enjoying lower costs. This would be manifest, for example, by greater variability in individual growth rate among some individuals than among others. To the extent that genetic heterozygosity is associated with differences in maintenance costs, more homozygous individuals would exhibit greater variance in individual

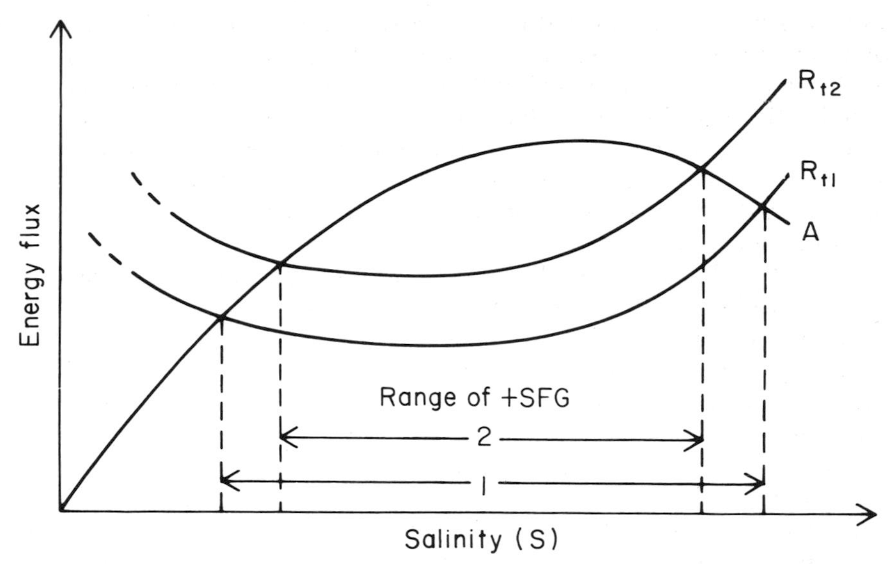

Figure 6. Proposed relationship between environmental salinity and energy flux (arbitrary units). Absorbed ration (A) is maximized at optimal salinity. Maintenance costs are minimum at the same point but increase at both higher and lower salinities; these maintenance costs can differ between individuals leading to different total metabolic rates (R_{t1}, R_{t2}). Lower maintenance is then associated with positive SFG over a greater salinity range.

growth rate than more heterozygous individuals (cf. Zouros *et al.*, 1980; Koehn & Gaffney, 1984). More homozygous individuals would suffer negative energy balance in environments that deviate less from the optimum than more heterozygous individuals. Growth, growth rate and net energy balance, then, are all a function of the environment, but the specific consequence of a particular environment on energy balance will greatly depend also on individual genotype.

We have emphasized growth as a reflection of net energy balance. However, growth is only one of several measures of performance that depend on energy balance. The genetic correlates (i.e. heterozygosity) that have been demonstrated for viability (Zouros *et al.*, 1983), fecundity (Rodhouse *et al.*, 1986), developmental stability (Lerner, 1954; Mitton & Koehn, 1985) and scope for activity (Mitton *et al.*, 1986) can also be treated as we have treated growth and the underlying energetics.

ACKNOWLEDGEMENTS

We gratefully acknowledge Anthony J. S. Hawkins for stimulating discussions and Walter J. Diehl for sharing unpublished data. This is contribution No. 676 from the Program in Ecology and Evolution, State University of New York, Stony Brook, New York, 11794.

REFERENCES

BAYNE, B. L., 1985. Responses to environmental stress: tolerance, resistance and adaptation. In J. S. Gray & M. E. Christiansen (Eds), *Marine Biology of Polar Regions and Effects of Stress on Marine Organisms:* 331–349. London: Wiley & Sons.

BAYNE, B. L. & NEWELL, R. C., 1983. Physiological energetics in marine molluscs. In A. S. M. Saleudden & K. M. Wilbur (Eds), *The Mollusca, 4:* 407–515. New York: Academic Press.
BAYNE, B. L., HAWKINS, A. J. S. & NAVARRO, E., 1987. Feeding and digestion by the mussel *Mytilus edulis* L. (Bivalvia: Mollusca) in mixtures of silt and algal cells at low concentrations. *Journal of Experimental Marine Biology and Ecology, 11:* 1–22.
BAYNE, B. L., THOMPSON, R. J. & WIDDOWS, J., 1973. Some effects of temperature and food on the rate of oxygen consumption by *Mytilus edulis.* In W. Wieser (Ed), *Effects of Temperature in Ectothermic Organisms:* 181–193. Berlin: Springer-Verlag.
BAYNE, B. L., HAWKINS, A. J. S., NAVARRO, E., IGLESIAS, I. P. & WORRALL, C. M., 1989. The effects of seston concentration on feeding, digestion and growth in the mussel *Mytilus edulis* L. (Bivalvia: Mollusca). *Marine Ecology Progress Series,* in press.
BRETT, J. R., 1958. Implications and assessments of environmental stress. In P. A. Larkin (Ed.), *Investigations of Fish-Power Problems:* 69–83. Vancouver: H. R. MacMillan Lectures in Fisheries, University of British Columbia.
BUSH, R. M., SMOUSE, P. E. & LEDIG, F. T., 1987. The fitness consequences of mutiple-locus heterozygosity: the relationship between heterozygosity and growth rate in pitch pine (*Pinus rigida* Mill.). *Evolution, 41:* 787–798.
DANZMANN, R. G. & FERGUSON, M. M., 1988. Developmental rates of heterozygous and homozygous rainbow trout reared at three temperatures. *Biochemical Genetics, 26:* 53–67.
DANZMANN, R. G., FERGUSON, M. M. & ALLENDORF, F. W., 1988. Heterozygosity and components of fitness in a strain of rainbow trout. *Biological Journal of the Linnean Society, 33:* 285–304.
DIEHL, W. J., 1989. Genetics of carbohydrate metabolism and growth in *Eisenia foetida* (Oligochatea: Lumbricidae). *Heredity,* in press.
DIEHL, W. J., & KOEHN, R. K., 1985. Multiple-locus heterozygosity, mortality, and growth in a cohort of *Mytilus edulis. Marine Biology, 88:* 265–271.
DIEHL, W. J., GAFFNEY, P. M., MCDONALD, J. H. & KOEHN, R. K., 1985. Relationship between weight standardized oxygen consumption and multiple-locus heterozygosity in the marine mussel *Mytilus edulis* L. (Mollusca). In P. Gibbs (Ed.), *Proceedings of the 19th European Marine Biology Symposium:* 531–536. Cambridge: Cambridge University Press.
DIEHL, W. J., GAFFNEY, P. M. & KOEHN, R. K., 1986. Physiological and genetic aspects of growth in the mussel *Mytilus edulis.* I. Oxygen consumption, growth, and weight loss. *Physiological Zoology, 59:* 201–211.
FRY, F. E. J., 1947. Effects of the environment on animal activity. *University of Toronto Studies in Biology, Series 55:* 1–62.
FUJIO, Y., 1982. A correlation of heterozygosity with growth rate in the Pacific oyster, *Crassostrea gigas. Tohoku Journal of Agricultural Research, 33:* 66–75.
GAFFNEY, P. M., 1986. Physiological genetics of growth in marine bivalves. Ph.D. Dissertation. Stony Brook: State University of New York.
GAFFNEY, P. M. & SCOTT, T., 1984. Genetic heterozygosity and production traits in natural and hatchery populations of bivalves. *Aquaculture, 42:* 289–302.
GARTON, D. W., 1984. Relationship between multiple-locus heterozygosity and physiological energetics of growth in the estuarine gastropod *Thais haemostoma. Physiological Zoology, 57:* 530–543.
GARTON, D. W., KOEHN, R. K. & SCOTT, T. M., 1984. Multiple-locus heterozygosity and physiological energetics of growth in the coot clam, *Mulinia lateralis,* from a natural population. *Genetics, 108:* 445–455.
GREEN, R. H., SINGH, S. M., HICKS, B. & McCUAIG, J. M., 1983. Antarctic intertidal population of *Macoma balthica* (Mollusca: Pelecypoda): genotypic and phenotypic components of population structure. *Canadian Journal of Fisheries and Aquatic Science, 40:* 1360–1371.
HAWKINS, A. J. S., 1985. Relationships between the synthesis and breakdown of protein, dietary absorption, and turnovers of nitrogen and carbon in the blue mussel *Mytilus edulis* L. *Oecologia, 66:* 42–129.
HAWKINS, A. J. S., BAYNE, B. L. & DAY, A. J., 1986. Protein turnover, physiological energetics, and heterozygosity in the blue mussel, *Mytilus edulis:* the basis of variable age-specific growth. *Proceedings of the Royal Society of London, Series B, 229:* 161–176.
HAWKINS, A. J. S., WIDDOWS, J. & BAYNE, B. L., 1989. The relevance of whole body protein metabolism to measured costs of maintenance and growth in *Mytilus edulis. Physiological Zoology,* in press.
HAWKINS, A. J. S., WILSON, I. A. & BAYNE, B. L., 1987. Thermal responses reflect protein turnover in *Mytilus edulis. Functional Ecology, 1:* 339–351.
HAZEL, J. R. & PROSSER, C. L., 1974. Molecular mechanisms of temperature compensation in poikilotherms. *Physiological Reviews, 54:* 620–677.
HOLLEY, M. E. & FOLTZ, D. W., 1987. Effects of multiple-locus heterozygosity and salinity on clearance rate in a brackish-water clam, *Rangia cuneata* (Sowerby). *Journal of Experimental Marine Biology and Ecology, 11:* 121–131.
KOEHN, R. K. & GAFFNEY, P. M., 1984. Genetic heterozygosity and growth rate in *Mytilus edulis. Marine Biology, 82:* 1–7.
KOEHN, R. K. & SHUMWAY, S. E., 1982. A genetic/physiological explanation for differential growth rate among individuals of the American oyster, *Crassostrea virginica* (Gmelin). *Marine Biology Letters, 3:* 35–42.
KOEHN, R. K., DIEHL, W. J. & SCOTT. T. M., 1988. The differential contribution by individual enzymes

of glycolysis and protein catabolism to the relationship between heterozygosity and growth rate in the coot clam, *Mulinia lateralis*. *Genetics, 118:* 121–130.

LERNER, I. M., 1954. *Genetic Homeostasis*. Edinburgh: Oliver and Boyd.

MITTON, J. B. & GRANT, M. C., 1984. Associations among protein heterozygosity, growth rate and developmental homeostasis. *Annual Review of Ecology and Systematics, 15:* 479–499.

MITTON, J. B. & KOEHN, R. K., 1985. Shell-shape variation in the blue mussel, *Mytilus edulis* L., and its association with enzyme heterozygosity. *Journal of Experimental Marine Biology and Ecology, 90:* 73–80.

MITTON, J. B., CAREY, C. & KOCHER, T. D., 1986. The relation of enzyme heterozygosity to standard and active oxygen consumption and body size of tiger salamanders. *Ambystoma tigrinum*. *Physiological Zoology, 59:* 574–582.

MOORE, M. N. & VIARENGO, A., 1987. Lysosomal membrane fragility and catabolism of cytosolic proteins: evidence for a direct relationship. *Experientia, 43:* 320–323.

MOORE, M. N., LIVINGSTON, D. R., WIDDOWS, J., LOWE, D. M. & PIPE, R. K., 1987. Molecular, cellular, and physiological effects of oil derived hydrocarbons on molluscs and their use in impact assessment. *Philosophical Transactions of the Royal Society, Series B, 316:* 603–623.

RODHOUSE, P. G., McDONALD, J. H., NEWELL, R. I. E. & KOEHN, R. K., 1986. Gamete production, somatic growth, and multiple-locus enzyme heterozygosity in *Mytilus edulis*. *Marine Biology, 90:* 209–214.

SCOTT, T. M. & KOEHN, R. K. 1989. The effect of environmental stress on the relationship of heterozygosity to growth rate in the coot clam, *Mulinia lateralis* (Say). *Journal of Experimental Marine Biology and Ecology,* in press.

SELYE, H., 1950. Stress and the general adaptation syndrome. *British Medical Journal, 1:* 1383–1392.

SINGH, S. M. & ZOUROS, E., 1978. Genetic variation associated with growth rate in the American oyster (*Crassostrea virginica*). *Evolution, 32:* 342–352.

SINGH, S. M. & ZOUROS, E., 1981. Genetics of growth rate in oysters, and its implications for aquaculture. *Canadian Journal of Cytogenetics and Cytology, 23:* 119–130.

STICKLE, W. B. & BAYNE, B. L., 1987. Energetics of the muricid gastropod *Thais* (*Nucella*) *lapillus* (L.). *Journal of Experimental Marine Biology and Ecology, 107:* 263–278.

VAN STEENBERGEN, E. J., 1987. Genetic variation of energy metabolism in mice. In N. W. A. Verstegen and A. M. Henken (Eds), *Energy Metabolism in Farm Animals:* 467–477. Dordrecht: Martinus Nijhoff.

WEDEMEYER, G. A. & McLEAY, E. J., 1981. Methods for determining the tolerance of fishes to environmental stressors. In A. D. Pickering (Ed.), *Stress and Fish:* 247–275. London: Academic Press.

WIDDOWS, J., DONKIN, P. & EVANS, S. V., 1987. Physiological responses of *Mytilus edulis* during chronic oil exposure and recovery. *Marine Environmental Research, 23:* 15–32.

WINBERG, G. C., 1956. Rate of metabolism and food requirements of fishes (Russian). BeloRussian State University, Minsk. (*Translations Series Fisheries Research Board of Canada*. 194.)

ZOUROS, E. & FOLTZ, D. W., 1987. The use of allelic isozyme variation for the study of heterosis. *Isozymes: Current Topics in Biological and Medical Research, 15:* 1–60.

ZOUROS, E., SINGH, S. M. & MILES, H. E., 1980. Growth rate in oysters, an over-dominant phenotype and its possible explanations. *Evolution, 34:* 856–867.

ZOUROS, E., SINGH, S. M., FOLTZ, D. W. & MALLET, A. L., 1983. Post-settlement viability in the American oyster (*Crassostrea virginica*): an overdominant phenotype. *Genetical Research, 41:* 259–270.

Biological Journal of the Linnean Society (1989), *37:* 173–181. With 5 figures

Proximate and ultimate responses to stress in biological systems

P. CALOW

Department of Animal and Plant Sciences, University of Sheffield, Sheffield S10 2TN

Environmental stress causes reductions in survival probability, growth rates and reproductive outputs. Under these circumstances, however, some genotypes may be less influenced by stress than others and these resistant forms will be favoured by stress acting as a selection pressure. Following Mayr *(Science, 134:* 1501–1506), these two effects can be described as proximate and ultimate respectively. The ultimate, evolutionary response can lead to direct adaptations that are either fixed or facultative, or to indirect life-history adaptations. Several attempts to classify stressful selection pressures are reviewed and relationships between these are considered. A model that potentially allows proximate and ultimate, direct and indirect adaptations to be treated within the same framework is described.

KEY WORDS:—Proximate, ultimate responses – fixed, facultative responses – life-history adaptation – resource-allocation model.

CONTENTS

INTRODUCTION

Environmental stress is a somewhat elusive term since it is both level and subject dependent. It is level dependent because a stress response at, for example, a molecular or cellular level, need not become manifest at an organismic one due to compensatory adjustments; i.e. due to homeostasis (Calow, 1976). On the other hand, individual organisms are such important operational units from both an ecological and evolutionary point of view that it is attractive to focus the definition on them (cf. Underwood, 1989). Selye's classical definition of stress, for example, specified a syndrome of physiological responses to environmental influences—alarm reaction, resistance, exhaustion—that influenced the health of the individual (Selye, 1973). Fry (1971), from a more ecological perspective, has linked environmental influences on the metabolism of organisms to effects on their 'scopes' for various metabolic activities, and in turn linked these to effects on ecological distribution and abundance.

0024–4066/89/050173+09 $03.00/0

A more wide-ranging definition of stress at the level of the individual organism, that extends that of Selye, is any environmental influence that impairs the structure and functioning of organisms such that their neoDarwinian fitness is reduced. The latter incorporates survival probability, developmental rate and fecundity and hence links responses at the level of individuals to parameters that influence the density of their populations and their future contribution to the gene pool. This approach has been followed by Bradshaw & Hardwick (1989), Hoffmann & Parsons (1989), Koehn & Bayne (1989), Nisbet *et al.* (1989) and Sibly & Calow (1989).

Following Mayr (1961), the immediate effect of an environmental stress can be referred to as a proximate response. But under these circumstances it is conceivable that resistant genotypes are favoured so that the ultimate response (again *sensu* Mayr, 1961) is the evolution of stress tolerance. This is why the impact of stress is also subject dependent (above). It is then debatable if tolerant genotypes or species can be described as being stressed. Even so they are adapted to resist stress, and it becomes interesting to consider the extent to which common suites of traits evolve under the same kinds of stress (Grime, 1989; cf. Bradshaw & Hardwick, 1989) or the extent to which stress can be categorized by reference to the evolution of common adaptations.

PROXIMATE RESPONSES TO STRESS

To be proximately stressful, an environmental variable must either increase or decrease from 'normal', and cause some change in organism and population (Underwood, 1989). However, the environmental stress may not be as straightforward as this, for it may not be a continuous variable (i.e. not a press but a pulse stress—see Underwood, 1989). Then it becomes important to characterize environmental stresses in terms of their amplitude, duration and frequency of occurrence. Clearly, a high amplitude stress applied infrequently could, in principle, have less impact than a lower amplitude version of the same stress applied more frequently (Underwood, 1989).

At an individual level, the stress will be registered as a reduction in survival and/or developmental (growth) rate and/or fecundity (above). At a lower level, it will have caused the structure and functioning of cellular and molecular systems to become impaired (Koehn & Bayne, 1989). At a higher level it will cause a change in population density (Underwood, 1989) and in the structural and functional attributes of communities—e.g. in diversity, dominance, primary production and community respiration (Gray, 1989; Rapport, 1989: table 1).

One way of attempting to understand a stress response is by describing the response that occurs at a particular level in this hierarchy. However, there are some difficulties associated with such a description and in identifying cause and effects in complex field situations (e.g. problems of pseudoreplication in space and time discussed by Underwood, 1989). So there is a trend towards investigating the effects of stresses by experimental manipulations in laboratories, quasi-field (enclosures of various kinds) and true field situations. The aim is to precisely control the stress input and to monitor the resulting output in terms of individual, population and community responses.

Yet there are still a number of problems in using this approach as a basis for understanding and prediction. First, there is the so-called black-box problem

(Calow, 1976); for one input there should be a particular output, but in complex systems a particular output may be brought about by more than one input. So the system to which the experimental perturbation is applied must be carefully controlled. Moreover, the input/output relationship for particular kinds of input and system need not be applicable to other systems. Thus generalization from particular experimental circumstances can also suffer from the problems of induction noted by Underwood (1989); see also Maltby & Calow (1989).

Sound predictions can only be based upon mechanistic understanding of the interaction between system and input—and this usually entails representing the system of interest in terms of machanisms at lower levels in the hierarchy—individuals in terms of physiology, populations in terms of individuals and so on. This is the so-called reductionist programme and has been advocated by Nisbet *et al.* (1989). It raises questions on how far this can be taken and the extent to which novel properties, e.g. due to interactions between individuals of the same or different species, appear at higher levels. The extent to which the physiological properties of individuals dominate the influence of ecological interactions between individuals in population dynamics would seem to be a matter for empirical exploration.

ULTIMATE RESPONSES TO STRESS

Figure 1 illustrates a simple life history. A propagule develops over a juvenile period and the organism breeds repeatedly over an adult period; survival probabilities and times over appropriate periods are represented respectively by

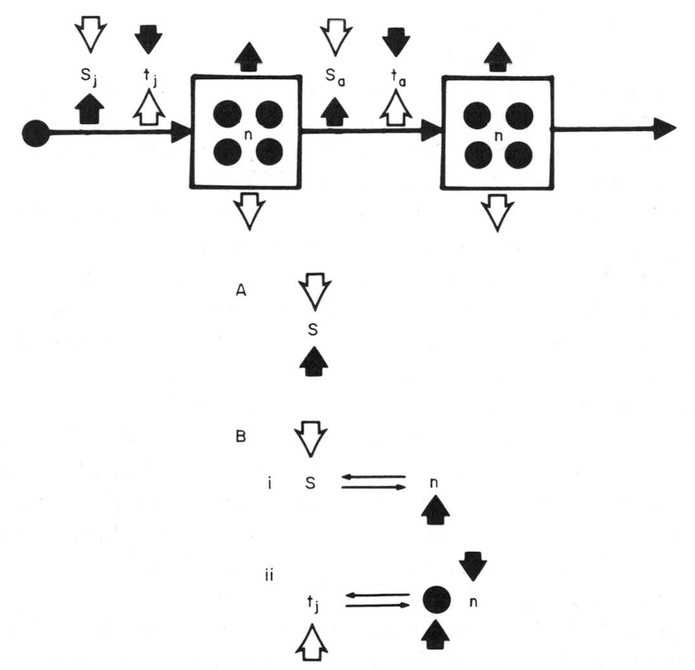

Figure 1. A diagrammatic representation of a life cycle. Spots = propagules; S = survival probability; t = time between developmental events; n = fecundity; subscript j = juvenile period; subscript a = adult period; solid arrow is natural selection; open arrow is stress. A, Represents a direct effect of selection; B, represents indirect effects of selection—see text. Modified from Maltby *et al.* (1987).

S and t and reproductive output by n. Natural selection will favour gene-determined traits that maximize S and n and minimize t, and stress, as suggested above, can be considered to have the reverse effects. It follows that stress can act as a selection pressure so that those traits that are least affected or not affected at all by the stress will spread at the expense of others leading to the evolution of tolerance. In principle there are at least two kinds of adaptation that can ensue (Maltby *et al.*, 1987).

A. Direct effects—in which there is genetic variance in the fitness component influenced by the stress; for example, if survivorship is impaired, some genotypes, possibly because of physiological mechanisms, are less susceptible than others. The trait so favoured may be fixed or facultative (Bradshaw & Hardwick, 1989), and in the latter case may be reversibly or irreversibly plastic (see table 4 in Bradshaw & Hardwick, 1989). These alternatives are represented in Fig. 2. The plasticity of the response is likely to depend upon the frequency of exposure and the cost of tolerance (see below).

B. Indirect effects—follow from fitness being a complex term (consisting of S, n, t and each of these being decomposable into age-specific components) and from some components being more susceptible to particular stresses than others. Because of the depression of one component of fitness, natural selection may favour genotypes that enhance other components of fitness. Two possibilities are illustrated in Fig. 1:

(i) is a situation where impaired survival favours enhanced investment in reproduction (in fact the situation is more complex since the age-specific effects have to be taken into account; Sibly & Calow, 1986).

(ii) is a situation where impaired growth rate favours a larger starting size because, other things being equal, this will shorten the time needed to reach an adult size. With a fixed investment in reproduction this may also cause reduced fecundity.

It seems likely that the extent to which indirect effects occur rather than direct ones will depend upon relative variance in different components of fitness and their ability to respond to stress as a selection pressure. The situation is complicated, though, because variance in one fitness component (say S) can be correlated (e.g. trade-off) with that of others (say n and t); as argued by Sibly &

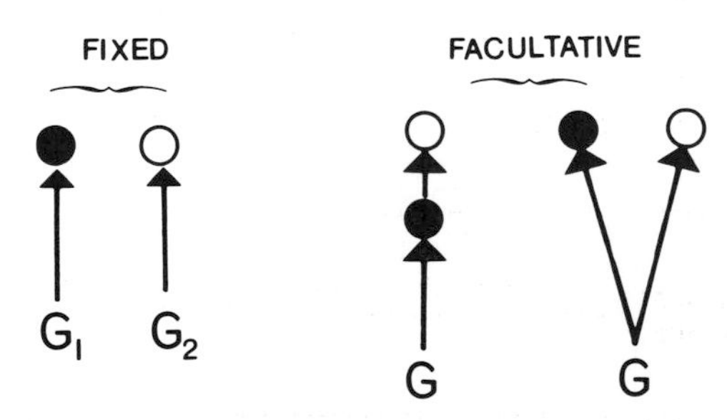

Figure 2. Fixed and facultative responses to selection. G represents genotypes and spots phenotypes. For fixed responses, either G_1 or G_2 is favoured, each giving a different phenotype. For facultative responses, different phenotypes can arise from the same genotype. The case on the left represents a situation where one or other phenotype might be induced depending on conditions and induction can be reversible. This might be described as a physiological response. In the case on the right, one of two phenotypes might arise dependent on conditions, but once induced the change is fixed. This might be described as a developmental response.

Calow (1989)—so direct effects could have implications for indirect ones and *vice versa*.

It follows from this that the effects of stress can probably be most effectively classified according to their effects on fitness components; i.e. upon S, n and t and their age-specific subcomponents. Hence in principle, the effects of stress and their classification ought to be represented in n-dimensional space. In practice, however, it has been more tractable to erect 2-(sometimes 3-) dimensional classifications that either represent distillations of the n-dimensional ones or focus on what are thought to be particularly important dimensions.

A series of such classifications is summarized in Fig. 3 together with a very simplified representation of some of the predicted traits associated with different selection pressures, from high to low stress. Given that these are all abstractions from a more complex adaptive space, it is not surprising that there should only be rough equivalence between them. For example, the mortality rate of Sibly & Calow (1989) is not exactly equivalent to their **S** (Sibly & Calow, 1985). The mortality rate is measured over a specified period of the life history, whereas **S** is a ratio of juvenile to adult survivorship (at least in absorption costing organisms; *sensu* Sibly & Calow, 1984).

Moreover, some of the components emphasize the characteristics of the environment (habitat templet; Southwood, 1977) rather than the impact on the organisms themselves. Yet, the implication must be that the habitat templet has its impact by influencing S, n, and t.

It is quite clear from these representations that environmental stress may shift systems around in more than one way within these matrices. The different forms

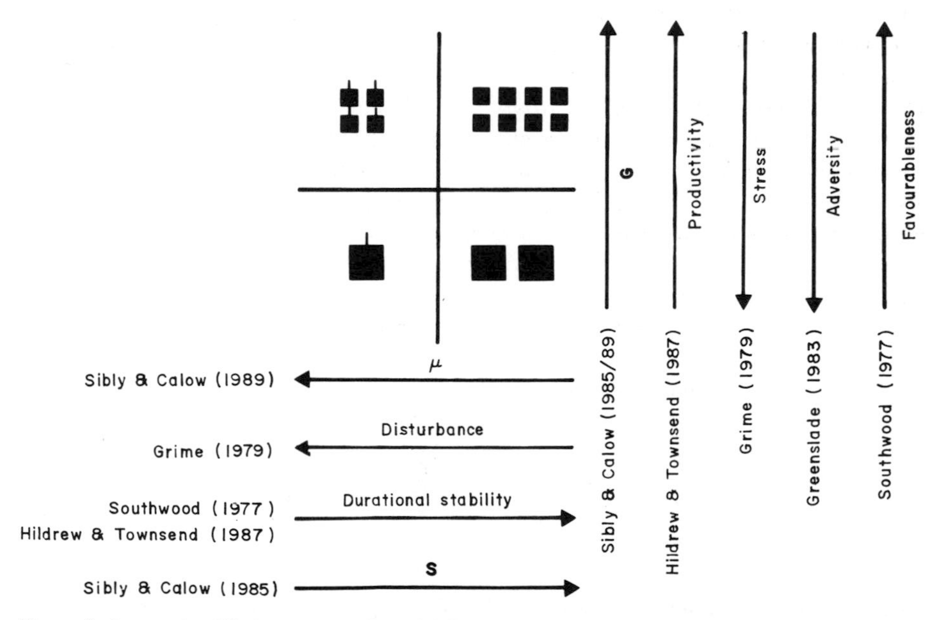

Figure 3. A very simplified representation of different classifications of selection pressures and their rough equivalence. Arrows representing each variable move from 'low' to 'high'. The blocks represent the kinds of response in reproductive effort (total area of boxes) and propagule size (area of individual boxes) that the classifications predict. The 'spines' emerging from the boxes to the left are intended to represent investment in defence. For a more detailed classification see Southwood (1988).

of stress may therefore be labelled according to the dominant axes along which they cause change; e.g. μ- and **G**-stress (Sibly & Calow, 1989), stress and disturbance (Grime, 1979). These may turn out to be just semantic matters, though, and the main consideration must be how the environmental stress influences S, n and t.

It is also important that the components of the classification should be capable of being precisely quantified. This is more difficult with the habitat descriptors than it is with the life-history descriptors, although, for the latter, it has to be conceded that organisms are used as probes of their own environments and this can lead to some problems (Sibly & Calow, 1986: 133).

An example of testing predictions about the effects of stress from a two-dimensional habitat classification

This example concentrates on the **S/G** classification of Sibly & Calow (1985) which focuses on indirect, ultimate effects of stress. As illustrated in Fig. 3, increasing **S** (juvenile to adult survivorship) favours increased investment in fecundity and *vice versa*. Similarly, reduced growth rate (**G**) favours a bigger starting size and *vice versa*. These predictions are derived rigorously from life-history theory and may not be mutually exclusive of other predictions that follow from other aspects of the theory (above).

Maltby *et al.* (1987) reported on a study of populations of the freshwater isopod, *Asellus aquaticus*, upstream and downstream of an effluent from an old coal mine (rich in acid and heavy metals). Laboratory observations indicated that juveniles were more sensitive to low pH and heavy metals than adults. Hence the effluent should reduce **S**; possibly a general outcome for this, but not necessarily all kinds of stress—for example, stress from low PO_2 (organic loading) could favour small juveniles (with large respiratory surfaces per unit volume). Heavy metal and acid stresses are likely to reduce **G**. Hence, with time, this kind of stress is likely to favour genotypes that make a diminished investment in fecundity but a larger investment per individual offspring.

From life-table studies and individual growth analysis for both populations it was possible to quantify **S** and **G** *in situ* (Fig. 4A) and as expected the pollution did shift individuals in the downstream population to the 'bottom left' of the matrix. *A. aquaticus* broods its young and so it was straightforward to measure

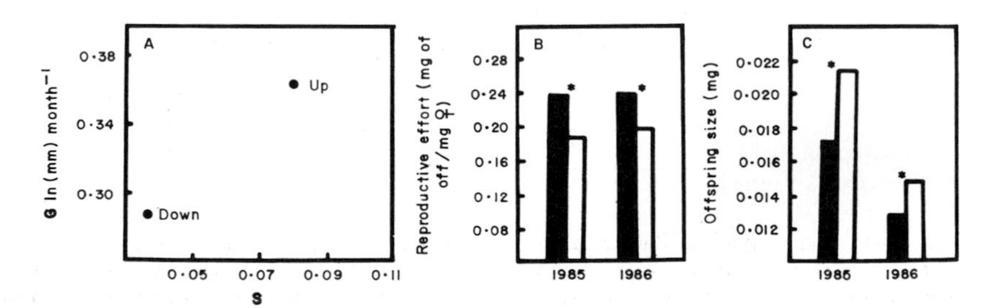

Figure 4. Testing predictions summarized in Fig. 3 on populations of *Asellus aquaticus*. A, Population locations upstream and downstream of a pollution source in the **S/G** matrix of Sibly & Calow (1985); B and C respectively show reproductive effort and neonate size in upstream (solid) and downstream (open) populations. Asterisks denote significant differences. After Maltby *et al.* (1987).

offspring size (mg dry weight per individual) and reproductive investment (effort) as mg dry weight of total brood per mg dwt per parent. As expected, reproductive effort was significantly lower and, more interestingly, offspring size significantly greater in animals from the polluted site (Figs. 4B, C). This difference persisted at the two sites for two years although, for reasons that are unclear, there was a difference between years at both sites. Of course, such responses could be due to ecological rather than evolutionary effects (Sibly & Calow, 1987), but Maltby (unpublished) has demonstrated that the differences persist after breeding in common laboratory conditions over several generations and this suggests that there is a genetic component to them.

This example illustrates how the components of the **S/G** classification can be quantified for natural populations and how predictions concerning stress effects can be investigated.

SYNTHESIS OF PROXIMATE AND ULTIMATE EFFECTS?

For reasons already noted (p. 173), a synthesis of proximate and ultimate effects is likely to be most appropriate at the individual level.

Figure 5, that models the organism as a resource acquiring and allocation system, is likely to prove useful. An acquired input, from feeding or photosynthesis, of resource (say energy) is allocated between catabolic and anabolic processes. The catabolic processes break down resources to yield energy for physical and chemical work (respiratory metabolism). The anabolic processes are concerned with the construction of somatic (growth) and reproductive (fecundity) biomass.

The molecular and cellular processes behind stress resistance are driven by and hence can be linked to the catabolic processes (Koehn & Bayne, 1989; but

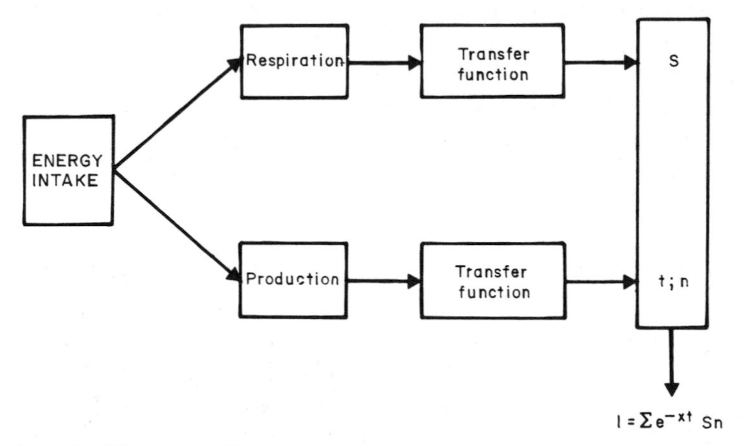

Figure 5. A synthesizing model. The organism is represented as a system that takes in resource (energy) and partitions it between respiration and production. The former generates power that is used in mortality avoidance (predator escape, combat of disease, resistance of stress) and hence should be functionally related to survival probability (S). Production generates biomass for growth and reproduction and hence should be functionally related to developmental rate and reproductive output (t, n). S, n, t can be brought together in the Euler-Lotka equation (of which there is a simple representation) to define population dynamics (x = intrinsic rate of increase) and fitness (x = Darwinian fitness).

see Hoffmann & Parsons, 1989). On the other hand, the allocations to catabolism and somatic and reproductive anabolism can, in principle, be translated into effects on survival probability (S), growth and hence time between developmental events (t) and reproductive investment (n). Hence, though the molecular, cellular and physiological traits associated with stress resistance may be very diverse (Bradshaw & Hardwick, 1989) they can, at least in principle, be related to general metabolic properties of organisms that can be linked in turn to demographic variables that influence population dynamics. For example, using the Euler-Lotka equation (Sibly & Calow, 1986) S, n and t can be related to the rate of population growth. Moreover, it is clear that if the input of resources to the organism remains relatively fixed, then any shift in allocation to one component of metabolism must be at the expense of allocations to other components. This provides a physiological basis for the trade-off models (Sibly & Calow, 1986, 1989).

It follows from this line of reasoning that physiological processes can be related explicitly and rigorously, if not straightforwardly, to population dynamics (Metz & Diekmann, 1986). Of course this approach ignores the possibility of interaction between individuals having an important effect at the population level; for example, a stress could influence S by influencing competition, predation or both rather than directly through the physiological processes described above. The importance of these interactive effects, as compared with influences on S, n, t through physiological modifications, is probably an empirical question.

At the same time, the allocation pathways must be enzyme modulated and hence ultimately gene controlled—something explored experimentally by Koehn & Bayne (1989). Hence, different genotypes will code for different metabolic strategies and will therefore be associated with different combinations of S, n and t. This then specifies the rate of spread of the genotype through the population—again expressible as the exponent in the Euler-Lotka equation—and the latter now precisely defines the neoDarwinian fitness of the genotype (Sibly, 1989).

Of course, the extent to which adaptation is possible depends upon the extent to which there is genetic variance for the trait in a population (Bradshaw & Hardwick, 1989). Amongst other things, this will depend upon physiological constraints. As an environmental stress intensifies it becomes more and more difficult for any genotype to cope. This is one reason why there are fewer species in these environments (Gray, 1989; Rapport, 1989).

It should also be noted that it is as energy transformers that organisms/populations 'plug into' ecosystem function, so that the kind of approach summarized in Fig. 5 also provides some opportunity for linking to community functional responses.

REFERENCES

BRADSHAW, A. D. & HARDWICK, K., 1989. Evolution and stress—genotypic and phenotypic components. *Biological Journal of the Linnean Society, 37:* 137–155.

CALOW, P., 1976. *Biological Machines. A Cybernetic Approach to Life.* London: Edward Arnold.

FRY, F. E. J., 1971. The effect of environmental factors on the physiology of fish. In W. S. Hoar & D. J. Randall (Eds), *Physiology of Fishes vol 1:* 1–98. London & New York: Academic Press.

GRAY, J. S., 1989. Effects of environmental stress on species rich assemblages. *Biological Journal of the Linnean Society, 37:* 19–32.

GREENSLADE, P. J. M., 1983. Adversity selection and the habitat templet. *American Naturalist, 122:* 352–365.

GRIME, P., 1979. *Plant Strategies and Vegetation Processes*. Chichester: John Wiley.

GRIME, P., 1989. The stress debate: symptom of impending synthesis? *Biological Journal of the Linnean Society, 37:* 3–17.

HILDREW, A. G. & TOWNSEND, C. R., 1987. Organization in freshwater benthic communities. In J. H. R. Gee & P. S. Giller (Eds), *Organisation of Communities Past and Present:* 347–377. Oxford: Blackwell Scientific Publications.

HOFFMANN, A. A. & PARSONS, P. A., 1989. An integrated approach to environmental stress tolerance and life-history variation: desiccation tolerance in *Drosophila. Biological Journal of the Linnean Society, 37:* 117–136.

KOEHN, R. K. & BAYNE, B. L., 1989. Towards a physiological and genetical understanding of the energetics of the stress response. *Biological Journal of the Linnean Society, 37:* 157–171.

MALTBY, L. & CALOW, P., 1989. The application of bioassays in the resolution of environmental problems; past, present and future. *Hydrobiologia,* in press.

MALTBY, L., CALOW, P., COSGROVE, M. & PINDAR, L., 1987. Adaptation to acidification in aquatic invertebrates; speculation and preliminary observations. *Annales de la Société Royale Zoologique de Belgique, 117 (Suppl. No. 1):* 105–115.

MAYR, E., 1961. Cause and effect in biology. *Science, 134:* 1501–1506.

METZ, J. A. J. & DIEKMANN, O., 1986. *The Dynamics of Physiologically Structured Populations*. Heidelberg: Springer-Verlag.

NISBET, R. M., GURNEY, W. S. C., MURDOCH, W. W. & McCAULEY, E., 1989. Structural population models: a tool for linking effects at individual and population level. *Biological Journal of the Linnean Society, 37:* 79–99.

RAPPORT, D. J., 1989. Symptoms of pathology in the Gulf of Bothnia (Baltic Sea): ecosystem response to stress from human activity. *Biological Journal of the Linnean Society, 37:* 33–49.

SELYE, H., 1973. The evolution of the stress concept. *American Scientist, 61:* 629–699.

SIBLY, R., 1989. What evolution maximizes. *Functional Ecology,* in press.

SIBLY, R. & CALOW, P., 1984. Direct and absorption costing in the evolution of life cycles. *Journal of Theoretical Biology, 111:* 463–473.

SIBLY, R. & CALOW, P., 1985. Classification of habitats by selection pressures; a synthesis of life-cycle and r/K theory. In R. M. Sibly & R. H. Smith (Eds), *Behavioural Ecology:* 75–90. Oxford: Blackwell Scientific Publications.

SIBLY, R. & CALOW, P., 1986. *Physiological Ecology of Animals. An Evolutionary Approach.* Oxford: Blackwell Scientific Publications.

SIBLY, R. & CALOW, P., 1987. Ecological compensation—a complication for testing life-history theory. *Journal of Theoretical Biology, 125:* 177–186.

SIBLY, R. M. & CALOW, P., 1989. A life cycle theory of responses to stress. *Biological Journal of the Linnean Society, 37:* 101–116.

SOUTHWOOD, T. R. E., 1977. Habitat, the templet for ecological strategies. *Journal of Animal Ecology, 46:* 337–365.

SOUTHWOOD, T. R. E., 1988. Tactics, strategies and templets. *Oikos, 52:* 3–18.

UNDERWOOD, A. J., 1989. The analysis of stress in natural populations. *Biological Journal of the Linnean Society, 37:* 51–78.

INDEX

Compiled by Stanley Thorley
Abbreviations:
Fig.—Figure
Tab.—Table
*—keyword.